THE CHARACTER OF CATS

ALSO BY STEPHEN BUDIANSKY

NATURAL HISTORY

The Truth About Dogs

The Nature of Horses

If a Lion Could Talk

Nature's Keepers

The Covenant of the Wild

HISTORY

Battle of Wits: The Complete Story of Codebreaking in World War II

THE CHARACTER OF CATS

The Origins, Intelligence, Behaviour, and Stratagems of *Felis silvestris catus*

Stephen Budiansky

Weidenfeld & Nicolson
LONDON

First published in Great Britain in 2002
by Weidenfeld & Nicolson

First published in the USA in 2002 by VIKING, Penguin Putnam Inc.

A CIP catalogue record for this book
is available from the British Library.

ISBN 0 297 60748 0

Printed in Great Britain by
Butler & Tanner Ltd, Frome and London

Weidenfeld & Nicolson

The Orion Publishing Group Ltd
Orion House
5 Upper Saint Martin's Lane
London, WC2H 9EA

The Cat.
He walked by himself,
and all places were alike to him.

—RUDYARD KIPLING

CONTENTS

ILLUSTRATIONS

Plate I (right). The African wildcat, *Felis silvestris* lybica group

Plate II (below). Cat and kitten in a domestic setting in ancient Egypt, around 1275 B.C. The cat, sitting beneath a chair, sports a collar and a silver earring. This scene is part of an elaborate wall painting from the tomb of Ipuy, a sculptor in the reign of Ramses II.

Plate III (left). A tame cat springs at a pigeon in this Greek vase painting from the sixth century B.C.

Plate IV (below). Common coat colors in cats: (a) the wild-type striped, or "mackerel," tabby; (b) non-agouti, or solid color; (c) blotched tabby; (d) sex-linked orange

ONE

Cats Plot to Take Over the World, and Succeed

There are no search-and-rescue cats, guard cats, Seeing Eye cats, bomb-detecting cats, drug-sniffing cats, escaped-convict-tracking cats, sheep cats, sled cats, gun cats, obedience-trained cats, Frisbee-catching cats, or slipper fetching cats. This is a matter of considerable relief. To tell the scientific story of dogs is to risk bringing down the wrath of legions of myth-soaked animal lovers, saturated as they are with tales of canine derring-do, loyalty, and "unconditional love," whatever that means. No one has any illusions about cats. Cats are cats, and any real cat owner knows it. That constant fraction of the human race that stalwartly admires and enjoys the company of cats long ago realized that they had better accept cats on their own terms, for the cats would have it no other way.

Dog science, inevitably, is about shattering myths. Cat science, rather more happily, is about explaining mysteries.

Of mysteries there is no shortage. Cats, with their shining eyes and silent footfalls, have always eluded explanation. Throughout the several thousand years of shared history between cats and human beings, cats have been a source of wonder and unease, reverence and superstition. Needless to say, given that man in his natural state is a simple and impressionable being, a certain amount of this mystification is the product of nothing more than man's own overworked imagination. Primitive peoples who lacked cats were perfectly capable of finding mystery and magic in rocks, trees, blades of grass, and cargo pallets dropped from Allied bombers.

But, in fact, cats really *are* mysterious. The ambivalent and superstitious emotions that the cat has evoked over the centuries mirror well the ambivalent and paradoxical place the cat truly occupies in nature and in the world of humans. Cats defy most of the normal rules about how and why animals came to enter the company of humans. The behavior of the cat in its association with human society is extraordinarily varied and complex: adaptive and perverse, affectionate and wary, gregarious and reclusive, dependent and aloof. The intelligence of the cat is an amalgam of extremes, of hard-wired instinct and adaptive learning. Cats have spread over the world in the company of man faster than man himself ever did, all the while keeping one foot in the jungle. Cats are the least tamed and the most successful of domestic species, the least altered within but the most changed in circumstance without.

So these mysteries are real—they are the product of nature, not merely our superstitious or ignorant imaginations—but even so they are our own doing in a way, because until recently science has ignored cats. The domestic cat's wild counterpart, the

European, African, and Asiatic wildcat *Felis silvestris,* is among the least studied of wild felines. It is a small, elusive, mostly forest-dwelling animal, and scientists were not able to find out much about the behavior, ecology, and genetics of small, elusive felines until the tools of molecular genetics and radiotelemetry lately began to change things. There has been a degree of scientific snobbery at work, too. Real wildlife biologists don't study pussycats. They don their safari jackets, clamber aboard their Land Rovers, and plunge down some rough and foreboding dirt track in dangerous pursuit of lions and tigers and bears (oh my). The flawed but longstanding belief held by many zoologists and ethologists that domestic animals are all just a bunch of sappy degenerates unworthy of serious scientific scrutiny has not helped, either. So the kind of insights that only science can offer—to help us understand why cats do the things they do, how they perceive their universe, and how they came to share, with such remarkable success, our homes and lives and hearts—has been notably absent from the considerable literature of the cat.

On the other hand, domestic cats did figure prominently in early studies of intelligence and learning and psychology, largely because they were so readily available and so cooperative. And in part because of that foundation there has been a great deal of new research on cognition and brain perception and the neurochemistry of emotion involving cats in recent years. A newfound recognition that all domestic animals represent a vibrant evolutionary story of adaptation and change has also brought newfound and well-deserved attention to the cat from evolutionary biologists and conservation biologists. And perhaps most of all, there is a cadre of basic research scientists today in fields ranging from the neuroscience of vision to molecular genetics who simply like cats, and who are eager to apply the tools of their trades to understanding what makes them tick. It doesn't hurt that many genetic diseases in humans, including hemophilia, diabetes, and Tay-Sachs disease, also occur in cats and that more than twenty-five genes responsible for such inborn diseases have been found in cats. That gives cat genome research a practical payoff, of interest to the powers that be that dispense research grants. In doing this very practical medical research, however, a lot of other stuff comes tumbling out, for the genome of a species is not just a catalogue of ailments or even a blueprint for an organism but also a recorded history of that species, of its travels and fate over space and time.

Cat science is the biography of a species. It is an exploration of where cats came from and how they flourished in the company of man, how they changed and how they stayed the same; it is about their wants and needs, their thoughts and urges, their rationality and their perversity, their group mores and their individual distinctiveness. Like any good biography, it is a tale worth reading for its own sake, but it is also a story with a moral: Cats are not so much pets as fellow travelers, and we impose our hopes and wishes and expectations upon them to our peril. They have their own biological niche and destiny, their own rules of social interaction, their own ways of ordering and perceiving the world. Their astonishing adaptability has found them a place with us, but that one foot is ever in the jungle. Understanding the true nature of cats, with all that science has to offer, is enlightening to us, and good for cats.

The Curious Uniqueness of the Cat

The cat stands as an anomaly among domestic animals. Every other domestic species is social in the wild. The wild forebears and wild counterparts of dogs, sheep, cattle, horses, donkeys, goats, chickens, pigs, ducks, elephants, and camels all live in groups. As such, they all have well-developed rules for getting along with others. Principally, that consists of a system for effectively communicating threats or signifying submission so that constant bloody fights over food, mates, sleeping spots, and other shared resources can be avoided. It doesn't take a Ph.D. in animal psychology or even a $495 session with a dog whisperer to know what a dog means when he growls. For all the misunderstandings that still occur between humans and a social animal like the dog, most human dog owners succeed in getting at least their general intentions across most of the time because our species share a common language that carries over even to the novel circumstances domestication has thrust upon us. Speak in a rough tone to a dog and he cringes and drops the pair of underwear he's swiped from the laundry. This is not something that evolution ever anticipated the wolf's encountering, but the social tool kit with which dogs are equipped by virtue of their social ancestry is the adaptable stuff that binds our two species together.

The domestic cat's wild progenitor is, by contrast, completely solitary in its natural state. A female wildcat lives in an exclusive territory of about 2 square kilometers (¾ square mile, or about 500 acres), which she will apparently guard against intrusion by any other female. The territory of a male typically overlaps that of several females, but again almost no overlap occurs with members of his own sex. Wildcats interact cordially with others of their kind only for the brief time when males and females meet to mate, and during the first four or five months of life when the kittens of a litter are together with one another and with their mother.

Within the large cat family there do exist social species—lions most notably; also cheetahs, which sometimes form small same-sex groups; and jaguarundis, which have been reported to live in pairs over parts of their range. But the relevance of the behavior of these social cats to that of the domestic cat is just about nil. The big cats branched off from the evolutionary line that led to the domestic cat some 9 million years ago; by way of comparison, that was several million years before the chimpanzee and human lineages diverged. Beginning around 12 million years ago the felid family underwent an explosive diversification, giving rise to thirty-seven species that today cover the Earth's geographical and ecological spectrum—from Siberia to the Amazon, from Arctic tundra to tropical forest. This diversification produced a correspondingly wide range of physical and behavioral types among those thirty-seven species. The members of the cat family vary from about 1½ to 300 kilograms in weight (about 3 to 650 pounds); their coats vary from striped to spotted to black, long coated to short coated. Domestic cats unmistakably show some intensely social behaviors, not only with human beings but also with one another. But the roots of this behavior are distinctly strange and anomalous, for there is nothing in the cat's immediate ancestry to account for it. That a distant relation of the domestic

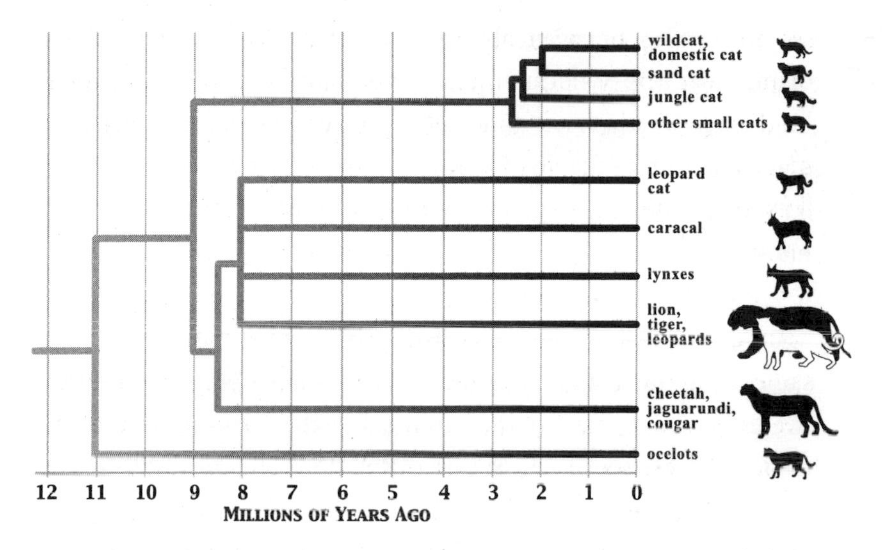

The cat family underwent an explosive diversification beginning about 12 millon years ago. Genetic researchers have reconstructed the cat evolutionary tree by analyzing DNA similarities between species.

cat like the lion shows social instincts does not get us anywhere. It makes no more sense to attribute the domestic cat's sociability to the lion than it would to argue that a domestic cat really ought to weigh a quarter of a ton and hunt zebras.

Dogs, cattle, sheep, goats, and just about all other domestic species show marked physical divergences from the wild type. If you show a trained expert a wolf skull and a dog skull, he will have little difficulty telling the two apart. But when zoologists, natural history museum curators, hunters, veterinarians, game wardens, and professional naturalists were asked to distinguish between specimens of the European wildcat and the domestic cat, they got it right only 61 percent of the time. That is not much better than flipping a coin, which would have yielded a 50 percent score. Except for the superficial alterations in coat color and coat length that have arisen in some domestic cats (traits that are almost surely a result of deliberate human selection), there is little in their anatomy that consistently distinguishes wildcats from domestic cats. European wildcats have larger brains as a proportion of body weight than do domestic cats; they have more retinal cones, the nerve cells responsible for color vision; and they have proportionately longer legs. Domestic cats also have longer intestines than do European wildcats, possibly a reflection of their adaptation to a more mixed and less carnivorous diet (you need longer guts to extract nutrients out of plant food). African and Asiatic wildcats probably share the European wildcats' proportionately larger brain and longer legs, but there is just too little data available on the other points. But even these distinctions between the wild and tame branches of *Felis silvestris* are subtle and certainly not apparent to the casual observer. Most people seeing an African wildcat (plate I) would think "pussycat." Even the normal wildcat coloration of striped, or "mackerel," tabby occurs frequently in domestic cats, whereas in other domestic species wild-type coloration (the dun-colored domestic horse is one example) is rare, or absent altogether.

Genetic measurements confirm the distance that most domestic species have drifted from wild populations. Domestic dogs and wolves show significant and consistent differences in certain stable DNA markers. Domestic horses have genetically diverged so far from the surviving wild type, Przewalski's horse, that they actually carry a different number of chromosomes. Enough change has occurred that dogs, horses, and most other domestic animals clearly qualify as biologically distinct species from their wild counterparts. But, again, with the cat there is almost no discernible difference between domestic and wild types. Gene sequences analyzed in domestic cats, African wildcats, and European wildcats differed from one another by only three to five nucleotide substitutions—that is, instances in which one chemical "letter" of a DNA sequence has been replaced by another. That degree of difference is so small as to fall within the noise level; a like degree of variation is routinely found from one domestic cat to another or from one African wildcat to another. The next closest genetic relative of the domestic cat, *Felis margarita,* the sand cat, differs by more than two times as much, ten nucleotide substitutions.

In taxonomy there is a continual struggle between "lumpers," who group different subpopulations together and, emphasizing their fundamental similarities, designate them as members of a single species and "splitters," who focus on the differences and give them each their own species name. In the past, the splitters were in ascendance in the cat world, and decreed all of the various populations of the wildcat/domestic cat to be separate and distinct species: *Felis catus,* the house cat; *Felis silvestris,* the European wildcat; *Felis lybica,* the African wildcat; and *Felis ornata,* the Asiatic wildcat. But there is now a growing consensus that it simply makes no sense to pretend that these are anything but members of one and the same species. All of the wild populations are indistinguishable from one another anatomically. (At one time it was thought that the European wildcats were significantly bigger and heavier than the African wildcats, but an "extensive series" of weight measurements proved this to be false; the European cats just looked bigger because of their thicker winter fur.) Genetic data suggest that significant gene flow among the African, Asiatic, and European populations occurred as recently as 10,000 to 15,000 years ago. By way of comparison, the major human races, undeniably members of the same species, first branched apart at least 40,000 years ago. All of the wildcat populations, along with the domestic cat, are thus now generally considered to be *Felis silvestris,* and many scientists are reluctant even to assert that these different branches amount to so much as distinct subspecies, let alone species; they refer to the lybica, silvestris, and ornata "groups" and leave it at that. Only the domestic cat is now generally recognized as deserving of its own subspecies designation, which is usually styled *Felis silvestris catus.*

Subtle variations in coat color do occur among the wildcat populations: All are basically mackerel tabby, but the European wildcats usually have more distinct tabby markings, the African wildcats have reddish or rust-brown fur on the backs of the ears, and the Asiatic wildcats have a more spotted patterning. But even these differences are not sharp or fast; they tend to vary on a continuous gradient across each population, further argument that all wildcats are really the same species.

In the past, some experts argued that the genetic and morphological similarities between domestic cats and wildcats were a consequence of cats having been among the most recently domesticated of domestic animals; there simply has not been enough time for major differences to emerge. Dogs go back at least 15,000 years and possibly much further; domestic sheep and goats appeared about 9,000 years ago; domestic cattle 7,000 years ago; domestic horses 6,000 years ago. By contrast, the first definitive historical evidence of domesticated cats—Egyptian paintings that depict cats in clearly domestic settings, such as sitting beneath a chair in a house (plate II)—dates from only about 1500 B.C. Unlike domesticated dogs, horses, sheep, goats, cattle, and camels, the cat is nowhere mentioned in the Hebrew Bible.

But there is at least some strongly suggestive evidence of a much earlier association between humans and cats. In the 1980s, a jawbone of *Felis silvestris* was excavated at a late Stone Age site at Khirokitia on the island of Cyprus, dating from 6000 B.C. There is no evidence that wild *Felis silvestris* ever existed on Cyprus or other Mediterranean islands, and it is thus almost certain that this cat, found in association with a human site, must have been domesticated, or at the very least deliberately carried from the mainland by human beings. If domestic cats go back that far, there should have been plenty of time for cats to acquire the genetic and morphological changes seen in the other major domestic species.

Perhaps the oddest thing that sets the cat apart from other domestic species is the curious fact that the cat's wild counterpart continues to thrive in large numbers. The wild types of most domestic species are extinct or on the verge of extinction. Przewalski's horse survives only in zoos and in artificially managed herds. All wild sheep species are critically endangered. The aurochs, the wild ancestor of domestic cattle, is completely extinct. The wild population of wolves numbers probably no more than 150,000 throughout the world. But *Felis silvestris*, practically alone, stands as a brilliant success as both a domestic animal and

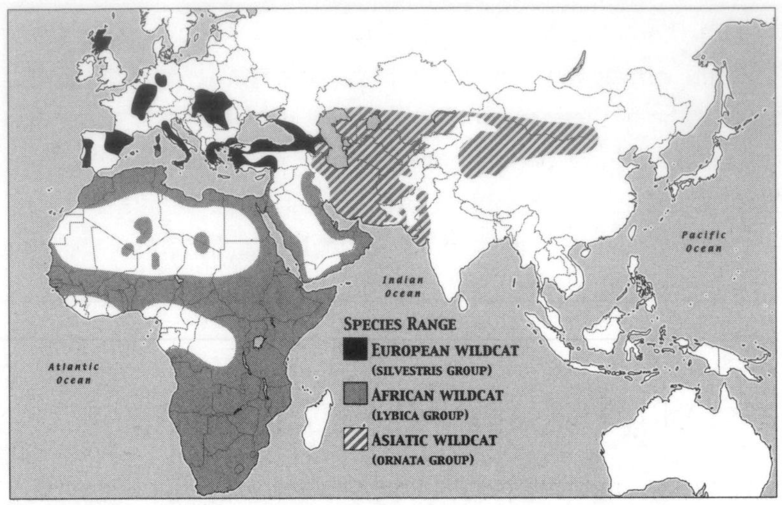

The wildcat, though currently threatened by interbreeding with domestic cats, remains widespread throughout Europe, Africa, and Asia.

a wild population. Although long trapped and hunted for their fur and persecuted as a threat to domestic poultry and game, wildcats have actually expanded their range in many parts of Western Europe since the Second World War, and remain widespread and common in much of Central Asia and Africa. Unlike most of the other thirty-six species of the cat family found in the world today, the wildcat is not considered endangered or threatened.

Though the degree to which wildcat populations have, over time, hybridized with domestic cats is a matter of considerable uncertainty, there is no doubt that huge populations of free-ranging cats—whether purely wild, wildcat × domestic-cat hybrids, or feral—exist in parallel to the owned, pet cat population throughout the world. (Feral cats are domestic cats, or the descendants of domestic cats, now living on their own in the wild.)

No one knows with any certainty how many wildcats there are in Europe, Africa, and Asia, but even a conservative estimate would have to be in the many millions. In the United States, which has no native wildcat population, there are an estimated 40 million feral cats (versus 75 million owned, pet cats). Unlike so many other domesticates, cats had no particular need to throw their lot in with man to survive. They still don't.

Deftly Defying the Rules of Domestication

One reason that wild populations of a species dwindle or disappear once the species becomes domesticated is that domestication was usually an evolutionary last resort for a species already in trouble. We commonly think of domestication as a consummately human act. Yet the evidence suggests just the opposite: Domestication was for the most part a natural process propelled by forces of climate change, geography, and evolution far bigger than anything man could engineer—or, for that matter, prevent.

At the end of the Ice Age, some 15,000 years ago, the climate was changing dramatically as the glaciers receded and the Earth warmed. The natural habitats of wild sheep, cattle, horses, and other soon-to-be-domesticated species were vanishing. And life with man, agricultural and village-dwelling man in particular, represented a new ecological niche ripe for exploitation. Where it is possible to trace the place of domestication for a particular species, it tended to be in areas where small remnant subpopulations, at the extreme end of their original range, were forced into proximity to human populations. That geographical isolation made possible the genetic isolation necessary for a new species, now selected for its compatibility with man, to emerge.

Much of that selection was probably self-selection, at least at first. Those individual animals who were more curious, less fearful, and more adaptable were more likely to approach human settlements to rob crops from the fields, or bones and food scraps from refuse piles, and so survive in this new niche—and more likely to pass those traits on to their offspring. To the extent that human actions shaped this initial process, by, for example, driving off or killing the more aggressive intruders but tolerating the friendly or cute ones, that, too, was probably more "natural" than "artificial," in the sense of not being deliberately directed toward a goal. People were just being people, and just being people is in itself a mighty selective force in the course of evolution. As much as we like to think of ourselves as having been terribly bright for having come up with the idea of capturing and breeding wild animals for our own ends, the evidence suggests it was in fact usually the animals who took the first step. Many species have found it in their interests to associate with man, even species we actively attempt to discourage (with notably little success) such as rats, mice, starlings, and pigeons, which have all spread throughout the world in the company of human beings, brilliantly exploiting our settlement and food disposal habits.

Many people have supposed that cats followed this same pattern: Taking the initiative themselves, they approached agricultural settlements to exploit the rich supply of rodents that man inadvertently began furnishing as soon as he learned to harvest and store grain. But there is a good argument that this conventional view of how cats and people came together is completely wrong. In a very nice paradox, it is those species that changed the most under domestication that are the most likely to have domesticated themselves, while it is species such as the cat, which have changed the least, that are actually most likely to have been deliberately captured and bred by man from the start. The biologist Juliet Clutton-Brock classified the cat as an "exploited captive" rather than a truly domesticated animal, and that may be a very apt definition.

The genetic and behavioral argument for why this should be goes like this: One hallmark of domestication is the suite of biological changes that rapidly occur, and occur all together, during the process of largely unconscious selection by man. Experiments with red foxes by the Russian biologist D. K. Belyaev demonstrated that by selecting for nothing but tameness, such as a cub's willingness to be approached and handled by a human being, it was possible in just five generations to produce a strain of foxes that had acquired the whole domestication package. These foxes had piebald coats and drooping ears, they wagged their tails and barked just like pet dogs, and they whined and begged for attention from familiar people. Many domestic animals, when compared to the wild types, seem to behave much more like juveniles than adults. They are playful throughout their lives; they do not develop the strong territorial or hunting or foraging instincts of their wild adult counterparts; they show a juvenilelike dependence on others for food and attention.

The crucial point is that the Russian scientists did not deliberately select for any of these particular traits. Rather, they selected only in a general way for tameness, much as ancient man might well have done, and done largely unconsciously, in his first encounters with future domesticates. All of these changes that the foxes exhibited—a loss of fear, a retention into adulthood of juvenile behaviors, a lifelong dependence, and even the physical traits of floppy ears and piebald coats—appear to be linked genetically. All seem to be the product of relatively small changes in the "master" genes that regulate the process of development from infancy to adulthood. Tweaking these results is a disruption and rearrangement of behaviors and physical traits so that juvenile forms are retained or entirely novel combinations emerge. It is striking how consistently this package of new features is present among domesticated species.

But not in the cat.

So, to recap: Cats were *not* in evolutionary trouble in the wild; they did *not* need to throw in their lot with man to survive; they did *not* undergo the rapid and automatic genetic transformation that broke down the barriers between the wild and the tame in the case of other wild beasts that became malleable and accommodating partners of man. Primitive man succeeded in taming wolves, cattle, sheep, and other true domesticates largely because these species had the inherent genetic potential to tame themselves genetically once people appeared in their environment. Cats refused to play this game.

The very absence in the cat of these usual genetic alterations that characterize the process of domestication tends to suggest that it would have taken some quite deliberate human initiative to breach the barrier. Domestication was a rare event, but this was an even rarer one. Coincidences of climate change, habitat loss, geographical location, and proximity to human settlement conspired to trigger a built-in genetic mechanism for rapid evolution in the domestic species that today are truly dependent upon man for their very existence and survival, as both species and individuals. This was a change of evolutionary sweep and historical majesty. It is not surprising that man's deliberate efforts to mimic it, thousands of years ago at the dawn of civilization, should have resulted in no more than a few genetic splashes in the pond, compared to evolution's genetic tidal wave.

Why Cats Fit In at All

Domestic cats, like camels, Asian elephants, deer, and other exploited captives, are thus probably not true domesticates. (It is nonetheless hard to avoid using the term "domestic cat," and I will continue to use it to refer to the populations of house cats, farm cats, and feral cats that biologically speaking constitute *Felis silvestris catus*.) But even if never truly domesticated, exploited captives such as the cat had to have been at least minimally tameable, and in other ways minimally compatible to living with man. After all, there are many species of mammals and birds that even with all the will in the world cannot be successfully kept in captivity, even by expert zookeepers—much less by millions of homeowners with little knowledge of animal behavior, ecology, or veterinary science.

So even when humans set out with determination and premeditation to keep a wild animal in captivity they are not likely to succeed on a large scale unless the animal meets some basic criteria. A fundamental rule is that a successful captive has to be a "generalist," that is, an animal that can eat a wide variety of foods. Most apartment dwellers would not find it very convenient to come up with a few hundred pounds of bamboo each week to feed a giant panda, for example.

More subtly, some species *are* just plain more tameable than others, and this is not something that always equates directly with being a social animal or a domesticated animal. All domesticated animals are tameable; but many tameable animals are not domesticable, and some tameable animals are not even naturally social animals. Within the cat family, both the cheetah and the caracal were once widely kept by the elite classes of the Near East and India as hunting companions. Cheetahs were kept in ancient Egypt, Sumeria, and Assyria, and in Mogul times in India; they were brought to the hunt often hooded, like a falcon, and then set loose when the prey was in sight. If they successfully caught the prey they were rewarded with a portion of the kill. One Mogul prince kept hundreds of cheetahs at one time; and the animals were said to be playful, loyal, and affectionate once tamed. The caracal, a long-eared, medium-sized (about 15-kilogram, or 30-pound) cat of Africa and Asia, is also easily tamed and has been used by hunters in Iran and India. It is the fastest cat of its size. Though it is not a social animal in its natural state, it, too, seems to adapt well to life with man. Neither the cheetah nor the caracal can be called a domestic species, and the caracal is not even a social species, yet both are tameable.

By contrast, people who have recently tried to keep as pets crosses between the domestic cat and the Asian leopard cat have had a pretty rough time of it. The leopard cat, *Felis bengalensis*, is almost exactly the same size as the domestic cat and is native to forests from Siberia to southeast Asia and India, much the same habitat that *Felis silvestris* is commonly found. Several leopard cats were kept by the National Institutes of Health for a gene-mapping project that involved crossing them with domestic cats (with which they are fully interfertile). When the project was done, several of the lab workers adopted the second-generation kittens, backcrosses that were three-quarter domestic cat and one-quarter leopard cat. They are beautiful animals and had been brought up with a lot of human handling, and they were certainly much tamer than their pure leopard cat grandparents. But one of the workers who adopted one of the cats soon asked to return it because the cat was *eating*—as in consuming, not merely chewing on—his leather shoes. Another one of the adoptees simply disappeared within his owner's house for long periods of time. She searched the entire house, but there was no trace of the cat. Finally she discovered that the cat had excavated a cave in the underside of her sofa and climbed in there to hide. The leopard cat and the wildcat both nest in hollow trees or rock crevices in their natural habitat, so it was not as if there was anything particular about the life history or ecology of the leopard cat that explains why leopard cat × domestic cat hybrids would become sofa diggers while pure domestic cats generally do not.

So what makes some wild species more tameable than others? The standard genetic process of domestication involves a rapid disruption of the process of development, which among other things results in a splitting up or rearrangement or just plain derailing of the highly developed adult behaviors that often are the source of trouble in captivity—things like territoriality, suspicion of novelty, or predatory behavior. But there was likely a similar, if much more subtle and drawn out, process of genetic "preadaptation" that took place in both the species that eventually became fully domesticated and those that made successful exploited captives. All of these species share a common history of having expanded their ranges dramatically during the Ice Age. As glaciers advanced and retreated, new territories and corridors opened up and then reclosed, and a number of land mammals, including sheep, camels, wolves, and cats took advantage of this to launch successive waves of colonization.

This process of colonization and adaptation to new habitats exerted an important selective force, for it put a premium on individuals that were adaptive, flexible generalists. There is some evidence, from studies of mountain sheep most notably, that species that were successful in expanding their range during the Ice Age underwent a genetic shift that in some ways anticipated the much more dramatic changes that later occurred under domestication. The mountain sheep species that are today the most distant geographically from their point of origin in North Africa show the greatest degree of what biologists term neoteny, the retention into adulthood of juvenile physical and behavioral traits. For example, the bighorn rams of the Rocky Mountains, which stand at the very end of the sheep's Ice Age hejira from North Africa to Siberia and finally across the Bering land bridge to North America, display far more juvenile behaviors (such as mounting ewes that are not in heat, or kicking as opposed to displaying their horns to threaten a rival ram) than do rams of the species they descended from, Stone's sheep of British Columbia.

The wildcat was also a great traveler. Starting from its Ice Age jumping-off point in Europe some 250,000 years ago, the wildcat spread in a series of migratory waves to Asia and Africa. The result is that *Felis silvestris* is today the most geographically widespread member of the cat family, and the most adaptable. Wildcats are found from the jungles of Africa to the northern woodlands of Scotland; they live in the deserts of Saudi Arabia and the Sahara, the semideserts of Central Asia, along seacoasts, in rocky scrublands, deciduous forests, and swamps. Across their range they are absent only from tropical rain forests, steppe grasslands, and places where deep snow remains on the ground for more than a hundred days of the year.

Of these far-flung wildcat subpopulations, the lybica group of Africa is the geographically most distant from its place of origin, the most widespread, and the most recent to arise. According to genetic analysis, the African wildcat achieved its definitive split from the European population only about twenty thousand years ago, just as the Ice Age was ending. And according to the theory of dispersal and neoteny, the African wildcat ought therefore to be the most juvenile and naturally tame. Despite the many very close similarities between the different wildcat branches, African wildcats do indeed show a consistent behavioral difference from the European branch. Simply, they *are* friendlier and tamer. European wildcats, even when hand reared, remain "fierce and intractable" according to one firsthand account. Crosses between house cats and European wildcats are equally wild. Andrew Kitchener, a zoologist at the National Museum of Scotland who is leading a major study on the wildcat, said that the one time he had the chance to meet European wildcat kittens in captivity the experience was distinctly "eerie." The four-week-old kittens "looked right through you as if you were not there," he said. They were totally silent and could not be encouraged to play or interact with people in any way. People who have raised European wildcats as pets report they are definitely "one-person" cats at best, and, like many wild animals kept in captivity, lose even that degree of tameness not long after reaching sexual maturity. Marianne Hartmann of the University of Zurich, who has studied European wildcats kept in large outdoor enclosures, says that some kittens start out extremely shy and hide whenever people approach, while others are almost tame. As they approach maturity, however, both types of kittens tend to become more similar in their behavior toward people: The shy cats lose their fear but retain their desire to be left alone, while the almost-tame cats retain their fearlessness but grow increasingly distant. They become, in Hartmann's words, bold and proud. Her adult wildcats will readily approach her but never allow themselves to be touched. (Hartmann says the adults in fact "behave as if they had the size of a leopard," and nonchalantly intimidate even large and fierce German shepherd dogs.) In other words, what keeps adult wildcats from forming close ties with people is not fear but aloofness.

Fearfulness toward people and emotional aloofness toward people thus appear to be separate components of the cat personality. Interestingly, experiments in which wolves were crossed with dogs have similarly suggested that these two components of behavior are separate and unlinked.

African wildcat × house cat crosses are, however, quite easily tamed; and even adult African wildcats captured from the wild begin to show tameness toward the people around them after a few weeks. The pure African wildcat kittens that the African zoologist Reay Smithers raised did, to be sure, remain rather wild in many ways. As adults they were very territorial toward one another and toward two Siamese house cats; and even as kittens, when punished for jumping up on the table, "instead of behaving like normal cats and flying through the door, would lay their ears back, spit, show their teeth, and strike back." Smithers's wildcats also wreaked havoc on the family's poultry, including ducks, bantams, and even some rather aggressive geese; only the peahen was able to keep the cats in line and avoid being swiftly killed.

Even so, these pure African wildcats were "superaffectionate" toward Smithers and his wife, purring and rubbing and demanding attention frequently, and they even formed a close attachment to the family dogs, rubbing themselves against the dogs' legs when they approached and curling up with them in front of the fire.

Though the African wildcat may never have tamed itself the way other domesticated animals did to become truly domesticated, it apparently did tame itself just enough so that it might be taken in by any reasonably adventurous human beings who came along.

Ancient Egypt

Trying to figure out who first took in a wildcat from the wild is a bit like trying to figure out who first invented the bookshelf or the practice of blowing one's nose in a handkerchief: The act of invention is not likely to have left much of a mark in the archaeological record. But whether or not the ancient Egyptians were actually the first to bring a cat in from the wild, there is no doubt that they did it on a scale unmatched by anyone who preceded them. Nor is there any doubt that the Egyptians perfectly fit the bill of being an

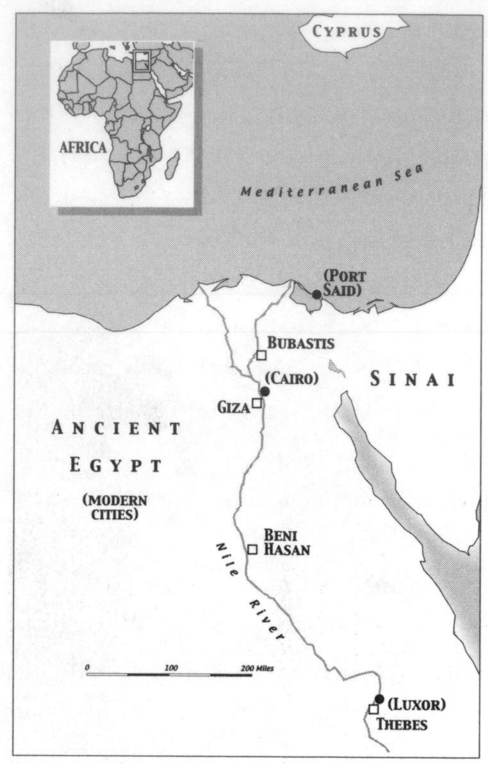

Sites associated with the Egyptian cat cult

extraordinarily adventurous people when it came to attempts to tame the wild. Tomb reliefs from ancient Egypt depict hyenas, ibexes, gazelles, and antelopes with collars around their necks and being fed—or perhaps force fed. Capturing and taming an already-disposed-to-be-tamed African wildcat could not have been a terribly difficult challenge for people who were willing to stick food down the throat of a hyena.

Although others may have got there first—and the circa 6000 B.C. cat jawbone from Cyprus certainly suggests that was the case—the Egyptians made the cat their own and, intentionally or not, spread it to the world. Certainly, no one ever accused the ancient Egyptians of thinking small, and in their devotion to the cat they came to be unrivaled in the ancient world. In 1888, a farmer plowing a field at Beni Hasan, a site on the Nile about halfway between Giza and Thebes, accidentally uncovered what surely ranks as the greatest cat cemetery of all time. In all, some 100,000 cat mummies were excavated from the sprawling catacomb, testimony to the great cult of the cat that held sway in ancient Egypt. (Creating an immediate glut on the nineteenth-century market for cat mummies, the entire haul was shipped off to Liverpool and, in a now famous story, sold for fertilizer at £4 a ton. There were 19 tons in all. Only a single specimen, now in the Natural History Museum in London, was preserved from the lot.)

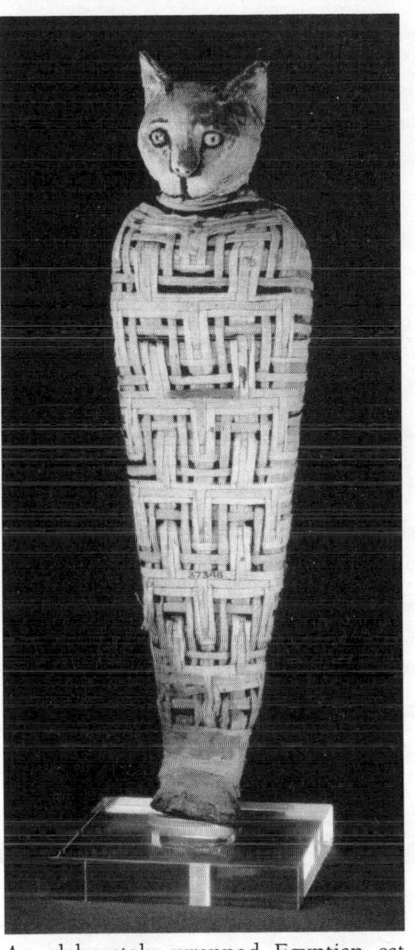

An elaborately wrapped Egyptian cat mummy from the Roman Period, after 30 B.C.

A few dozen cat mummies from other sites escaped that ignominious fate and survive to show the great care that was taken in preparing cats for burial. Many have elaborate wrappings in the intricate patterns that are typical of those used for the human notables of Egyptian society who could afford mummification. The heads of the cat mummies were covered with papier-mâché masks and then they were placed in mummy cases, sometimes accompanied by little shrew and mouse mummies to provide food for the afterlife.

The Egyptian cat goddess, Bastet, depicted as a woman with a feline head and kittens at her feet; she holds a ceremonial rattle and shield. From the fourth century B.C.

Sculpture and ornament likewise reflect the intense, popular Egyptian devotion to the cat, which, in the form of the goddess Bastet, was associated with fertility in the fields and the womb. Bastet received the particular devotion of young married women, who created a considerable market for bas-relief amulets bearing the image of the cat, which apparently served as fertility charms. Cats were also widely sculpted in every medium from gold and ivory to bronze, stone, wood, plaster, and mud.

The Egyptians' devotion to their pet cats, and cats in general, is also well documented from several historical sources dating from the first few centuries B.C. Herodotus, who visited the center of the Egyptian cat cult in Bubastis (modern Tell Basta, about 60 kilometers, or 40 miles, northeast of Cairo) around 450 B.C., described a huge temple containing a colossal statue of Bastet and thousands of live cats who were fed and cared for by the priests. Herodotus wrote of the mourning that members of the household went through if their pet cat died from natural causes; they shaved their eyebrows and, if they could afford it, took the dead animal to Bubastis to be embalmed and entombed by the priests. Diodorus Siculus, a Greek historian who traveled to Egypt in 60–57 B.C., related the more serious consequences that ensued if a cat's death was due to other than natural causes: "Whoever kills a cat in Egypt is condemned to death, whether he committed the crime deliberately or not. The people gather and kill him. An unfortunate Roman, who had accidentally killed a cat, could not be saved, either by King Ptolemy of Egypt or by the fear which Rome inspired."

This Egyptian abhorrence of harming a cat was apparently well known throughout the region, and was put to good use by an invading army of Persians in 525 B.C. Laying siege to Pelusium, Egypt's frontier fortress that stood about 30 kilometers (20 miles) east of modern Port Said, the Persian commander ordered his men to round up all the cats they could find. Just before launching their final assault, they set loose this horde of cats ahead of them. The Egyptian defenders of the city surrendered without firing a shot for fear of hitting a cat. The pharaoh Psamtek III was overthrown and a Persian dynasty installed over Egypt.

Contemplating this historical record, some have concluded that the cat held a place in the religion of the Egyptians akin to that of the sacred cow in Hinduism, and they have found in the Egyptians' reverence for the cat a deep attachment that mirrors the feelings of many modern-day cat owners. Among cat historians, ancient Egypt is often held up as a sort of golden age in which cats had it great—before the deluge of the Middle Ages and the Christian Church's vilification of the cat as familiar of witches and ally of the devil. It is also certainly tempting to see in the Egyptians' religious devotion an explanation for why they brought cats into their society in the first place.

Yet it is always dangerous to try to make too much sense of the religions of others, and the position and role of the cat in Egypt was probably rather more complex than it is often made out to be. The Egyptians' cat worship clearly did serve as a vector for the introduction of cats to mankind at large: By the first centuries B.C., the revered and coddled cats of Egypt had grown to a sizable population that soon erupted and spread throughout the Mediterranean and thence to the farthest reaches of the Roman Empire.

But it is more difficult to measure the degree to which religious feeling was behind the original introduction of the cat into Egyptian society many centuries earlier. The true answer may be "not much." For one thing, the Egyptian cult of the cat—as cat—appears to have been a quite late development. It only really began to take off after the first of the Libyan pharaohs, Shoshenq I of the Twenty-second Dynasty, who reigned from 935 B.C., made Bubastis his capital, and elevated what had been the small and local cult of Bastet to the status of the official deity of the kingdom. (Shoshenq I was the King Shishak of the Hebrew Bible, who, in addition to nationalizing the cat cult, invaded Jerusalem with an army of 1,200 chariots, 60,000 horsemen, and "innumerable" foot soldiers and ransacked the temple and palace of Solomon's son King Rehoboam.)

There is little evidence that cats were ever kept as pets in Egypt to any appreciable degree much before this time; the only common pet other than the dog in the Old and Middle Kingdoms (about 2700 to 1800 B.C.) was the monkey. And only dogs and cattle commonly received personal names at any time in ancient Egypt. Cats were nearly always called "cat." (In the Egyptian language, this was a word written with the three consonants *myw.* As in modern Hebrew and Arabic, the vowels in ancient Egyptian were not indicated in writing, so it is impossible to know exactly how this word was pronounced, but just about any vowels you can think of to stick in between those letters make it pretty obviously onomatopoetic.) The sole exception to the rule of calling a cat a cat that can be found in any ancient Egyptian writings was one cat named *njm,* which means "sweet."

The goddess Bastet can be traced back to these earlier periods, and she was frequently represented as a cat, or as a female human being with a feline head. But worshipping a cat goddess is very different from worshipping cats in general. Many Egyptian gods and goddesses were depicted in animal form from the earliest times without any accompanying veneration of the animal per se. It was not until the Late Period (beginning around 700 B.C.) that any of the Egyptian gods came to be identified literally with their earthly animal counterparts en masse. In earlier times, at most a single, chosen member of the species might be selected as a "divine herald" of the god. The most famous of these was the sacred bull Apis at Memphis, who was the herald of the god Ptah. But there was only one Apis at a time.

The wholesale veneration of all members of a species associated with a particular god or goddess, and the mass mummification and burial of animals, was a separate and much later cultic development; and it was actually part of the degeneration and death throes of the entire ancient Egyptian religion as it came under assault from the Persian, Greek, and Roman invaders. The cat was hardly alone in receiving this treatment. As the mythologist Donald Alexander Mackenzie observed:

> Egypt yearned for the glories of other days, and became an imitator of itself. Everything that was old became sacred; antiquarian knowledge was regarded as the essence of wisdom. . . . Animal worship was also carried to excess. Instead of regarding as sacred the representative of a particular species, the whole species was adored. Cats and rams, cows and birds, fishes and reptiles were worshipped wholesale and mummified.

Great cemeteries of crocodiles, birds, and gazelles matched those of the great cat cemeteries. There was also a good bit of sharp practice in this religious revival. X rays of fifty-five cat mummies show that beneath their impressive outer wrappings, many of them were a sort of morticians' and taxidermists' remainder bin. One mummy that looked outwardly like a cat actually consisted of a cat's skull perched atop a few pieces of human leg bone.

The X rays also reveal that most of the cats were almost certainly deliberately killed, probably by temple priests eager to supply the demand from people who wanted to purchase a mummy as an offering to the deity. Many of the cats had had their necks broken, and all but two of the fifty-five specimens were under two years of age when they met their end; about a third were less than four months old. None of this is consistent with their having died a natural death at the end of a long and venerated existence.

So it may be a mistake to think that the Egyptians brought cats into their society because they worshipped them as gods; it is probably more apt to say that the Egyptians were an adventurous and curious people who brought into their civilization all sorts of animals from the wild—from baboons to hyenas to antelopes—and cats were one species that took. The Egyptians worshipped cats as gods because they worshipped all sorts of things as gods, and cats were ready at hand.

On the other hand, the reverence that cats en masse did in time receive from the Egyptians, especially from the Late Period on, surely helped domestic cats to reach a critical population mass, and from that point there was no going back—either for cats or for mankind.

Cats Wild, Tame, and In-Between

The spread of domestic cats from Egypt was slow at first. For a long while Egypt actually forbade the export of cats, and Egyptian travelers apparently scooped up any cats they encountered in foreign lands and repatriated them. Evidence of domestic cats does not appear in Greece until about 500 B.C., in the form of a marble relief depicting an encounter between a dog and a cat, both leashed, while two seated men look on. A Greek vase from this same period shows an apparently tame cat leaping at a bird (plate III). No remains of cats were found at Pompeii, buried by the volcanic eruption of Mount Vesuvius in A.D. 79. But three centuries later cats had spread throughout the Roman Empire; impressions of cat feet appear on clay tiles from the fourth-century Roman town of Silchester in southern England.

The dispersal of cats from Egypt was accelerated by the collapse of the cult of Bastet, which had survived into the Christian era but was at last abolished by the decree of Emperor Theodisius I in the fourth century that banned paganism throughout the Roman world.

Cats were no doubt carried deliberately by human beings on their travels by foot or ship, and to that extent they were under human control. Yet cats continued to exist in many worlds simultaneously in a way that defied, or at least circumscribed, human agency. Unlike horses and cattle and sheep, whose matings could be readily controlled by human handlers and which were thus probably selected and bred for desirable traits from the earliest days of man's husbandry of them, there is no evidence that cats in ancient times were ever captively bred in any sense. Even today in the United States, less than 3 percent of owned cats are a "breed"; nearly all matings are random and determined by the cats in question, not by their owners. This is true even of the overwhelming majority of house cats which, until only very recently, were always free to come and go. Studies of present-day cat owners in America indicate that still some 40 percent of pet cats in urban and suburban areas are allowed outdoors, plus probably close to 100 percent of cats in rural areas, where three-fourths of residents report owning a cat. The mass of owned house cats and farm cats thus for millennia have led a kind of double life: tame, dependent upon, and often affectionate toward their nominal human owners, yet at the same time free living and involved in a complex cat social world beyond the four walls of their homes.

Simultaneously, a second large population of domestic cats sprung up around human settlements: stray pets, "neighborhood" cats, and feral cats—the descendants of domestic cats spread by man's travels, now living a largely wild existence. This large pool of free-roaming, unowned domestic cats equals or exceeds the owned cat population in many places even today and occupies a broad continuum of social relations with human beings. Many free-roaming cats receive regular handouts from people; one survey in Brooklyn found that about 20 percent of households regularly left out food for cats and in all provided considerably more food, in fact, than needed to support the entire estimated population of unowned cats in the city. Studies of free-roaming cats in Baltimore, Maryland; Rome; and Portsmouth, England, found that garbage was a major source of food as well for urban free-roaming cats. And about a third of pet cats are actually recruited from the free-roaming cat population, creating a continual interchange between the pool of owned and unowned cats. At the other end of the free-roaming spectrum are truly feral cats, which remain almost invisible to and largely wary of human beings, emerging only at night from hiding places in barns and cellars and dumps. But even they for the most part remain ultimately dependent on human activity for their survival, scavenging garbage or hunting rodents drawn to farm fields and grain bins.

The fact that almost all domestic cats—owned, stray, and feral alike—live where people live would have helped to keep them genetically isolated from the surrounding wildcat population. People have hunted wildcats from prehistoric times for their pelts, and farmers have traditionally viewed them as a threat to poultry; thus people tend to maintain a wildcat-free buffer around their settlements. Moreover, the territoriality of domestic toms would have served to fight off encroachment by wildcat toms seeking females in heat—and vice versa.

But the tameability of African wildcats to this day and the wide range of behaviors seen in domestic cats depending on how they are reared—from extremely affectionate in kittens brought up with a lot of social contact to downright vicious in many feral cats that never had much exposure to humans—suggest that much of the behavioral differences that separate domestic cats from their wild counterparts are not really genetic. This is very different from cattle, sheep, dogs, and most other truly domesticated species. Domestic cats do show one very dramatic behavioral change even from African wildcats, in that they will tolerate living in social groups, and even form them on their own. Feral domestic cats are often found in groups of many females living together in a colony surrounding a food source such as a dairy barn or a dump. The females in such groups will even share in the rearing of their young, allowing kittens who are not their own to nurse from them. Wildcats, even African wildcats, never exhibit such behaviors even when concentrated food sources are available. They are true loners, and tend to keep their distance from other adult wildcats.

The evidence intriguingly suggests that the gregariousness of domestic cats *is* inherited, but inherited culturally, not genetically. Kittens that are brought up in social groups, and who therefore become accustomed to interacting with unrelated adults from infancy, retain that propensity throughout their life; as adults they are willing to form bonds with other adults. Kittens brought up only by their mothers, with no social contact with strangers (cat or human), remain loners as adults.

Human beings thus arguably acted as an irreversible catalyst for a social revolution in the cat world. The subpopulation of African wildcats that was brought into contact with man was exposed to a social environment that taught their kittens to be gregarious as adults. From that point on the die was cast. Even subsequent generations that had little or no contact with humans would have maintained that cultural tradition, forming social groups of their own that then passed on the tradition of gregariousness to their feral offspring.

What really separates domestic cats from the wild may not be genes, but tradition.

TWO

Black Cats and Tabby Cats

Cats were brought into human society by a deliberate act of adventuresome and curious human beings; there they grew to a critical mass owing to the tolerance of a religious culture that in time came to venerate them en masse; thence they spread wherever man went, a vine circling the tree of humanity that always planted its own roots nearby, yet apart. For centuries humans exercised a small and almost unconscious influence on the direction that vine would grow and the form it would take.

Then as now, the overwhelming fact of the cat was its independence and ambiguity. The cat's ambiguous place in nature and human society was long reflected in the ambiguous place it occupied in man's feelings. The cat has evoked highly divergent emotions in the heart of humankind and has from time to time benefited, suffered, and been subtly altered by these feelings.

The divergence in human feelings toward the cat persists to this day. A survey of contemporary American attitudes toward the cat found that 17 percent of people expressed a dislike of cats, versus only 3 percent who expressed similar feelings about dogs. On the other hand, cats have since the 1980s surpassed dogs as the most popular pet in American households (there are currently some 75 million owned cats versus about 60 million dogs). The pet cat today is on the receiving end of a modern cult of worship that would put the Late Period Egyptians to shame, with everything from a multimillion-dollar-cat-toy industry to cat psychologists and even cat psychics who, for a fee, can tell you what your cat really feels about you. In the absence of the conscious and deliberate acts of breeders who took the fate of many domestic species in firm hand, the cat's fate in the company of man rested for centuries on unconscious cultural attitudes and imperfectly articulated beliefs.

A Brief History of Magic and Fear

By the eleventh century cats were considered so valuable as exterminators of rodents that in France they were often listed in wills and legacies. A mid-tenth-century Welsh legal code enumerated a complex schedule of prices for cats of various ages and rodent-killing experience, and specified fines for those who failed to take proper care of cats. The value of a good ratting cat was set equal to that of a fourteen-day foal, a six-month calf, or a weaned pig. If someone stole or killed a cat he was required to pay compensation to the owner in the amount of a ewe plus her lamb or a quantity of grain sufficient to cover the dead cat held by the tip of its tail so that its nose just touched the ground.

That was the good news. The bad news was that repeatedly in the late Middle Ages and early modern period in Europe there are recorded instances of cats being publicly tortured and slaughtered, and of accusations of witchcraft leveled against people who cared for or kept cats. On St. John's day in a number of French towns cats were hurled into bonfires or placed atop poles in the center of fires, a ceremony that in Metz lasted until the late eighteenth century. In witchcraft trials in the Old and New Worlds, old women who had cats came under particular suspicion. Tales collected by folklorists in the late nineteenth and early twentieth centuries are replete with incidents in which cats are associated with witches or bad luck. Many European folk beliefs hold that one should kill or maim any cat one encounters at night to ward off evil.

In what has become a de rigueur narrative in histories of the cat, the story goes that precisely because of its strong association with the ancient pagan "Mother Goddess" (in turn identified variously with such later pagan goddesses as Diana, Venus, and Freya), the cat underwent a dramatic reversal of fortune as the medieval church authorities sought to extirpate all survivals of pre-Christian cults. From a benevolent symbol of fertility and motherhood the cat became a literal symbol of the devil. Many scholars have gone further and read into this fate of the cat a sexual and psychological allegory, in which an animal notable for its promiscuity and highly vocal copulatory behavior was revered in the "feminine," nature-loving, egalitarian ur-religion of the Mother Goddess, and correspondingly reviled by the "patriarchal" and "misogynist" Christian Church. The metaphorical identification of cats with female lasciviousness is supposed to be the particular psychological key to this tale of medieval European vilification. James Serpell, who has written much in this vein, asserts flatly, "The unmitigated cruelty cats have received as a result of this metaphor speaks volumes about the sexual insecurities of European males."

All of this, of course, dovetails with lines of feminist scholarship popular in the world of academe these days, and also with the more-than-a-little self-serving insistence by certain practitioners of literary criticism that metaphor is reality, which conveniently saves one the bother of having to learn anything about science or history. But this oft-repeated tale of religious persecution is at best a gross exaggeration, a sweeping ideological theory anchored to a few selected facts. Both the extent of the supposed persecution of cats in medieval Europe, and its cause, have been greatly overstated.

Several of the assumptions that this analysis rests upon are simply historically inaccurate. True, cats have both fascinated and disturbed human beings for millennia. And no doubt, the cat's sexual behavior was part of what provoked both the interest and the unease. In French, for example, *la chatte* has the same vulgar slang meaning as "pussy" does in English; it is not unreasonable to read in both a metaphorical association of the cat with female promiscuity, or at least crude male notions of female promiscuity. So, too, has the cat's unworldly manner of moving silently on padded feet and seeming to appear and disappear mysteriously with glowing eyes been a source of unease among the superstitiously inclined. (Part of what makes the cat's movements so otherworldly seeming is that, unlike almost all other quadrupeds, cats walk and run with a sort of gliding motion in which the front and rear feet on one side advance together. Cats also have an extremely flexible backbone and, lacking a collarbone, can rotate their shoulders in almost any direction, all of which contributes to their remarkable litheness and ability to maneuver through small spaces.)

But the ambiguity in human feelings toward the cat was present all along, and so was the contradictory treatment of the cat at human hands. Cats were killed and cosseted in ancient times; they were killed and cosseted in medieval times. They were killed and cosseted by pagans; they were killed and cosseted by churchmen. Cats were adored by medieval monks and by Cardinal Richelieu himself, who kept dozens and left legacies for his favorites in his will; they were slaughtered by irreligious soldiers and irreligious and drunk printers' apprentices. They were witches' familiars even at the same time they were pets for millions. It is clearly a distortion of history to insist that there was some concerted anticat movement conducted under the auspices of the Christian Church in the Middle Ages, a sort of cat Inquisition.

It is an even greater distortion to put the cat forward as the sacrificial innocent trapped in a struggle between the Church and the ancient pagan worship of the Mother Goddess. For one thing, the whole notion that there ever *was* some ancient ur-religion centered on the worship of a Mother Goddess has been roundly debunked lately by scholars who have shown it to be the overenthusiastic invention of nineteenth-century folklorists and anthropologists (and the outright fabrications of one British civil servant and devotee of the occult named Gerald B. Gardner, who threw together a bunch of Masonic rites and other mumbo jumbo to found a modern witchcraft cult that he claimed was the direct descendant of an ancient goddess cult). Nor is there any evidence that the association of cats with witchcraft in folklore reflects any actual position that cats held in actual pagan beliefs in ancient times. It is impossible to tell just how far back the tales collected by folklorists go, but there is good reason to be skeptical. Court records show that during the witch trials that swept the New and Old Worlds in the late Middle Ages and early modern period, witches were accused of having many different familiars, of which the cat was only one; dogs, goats, pigs, rats, and other small animals also frequently figure in the charges. The association of cats with witches in many surviving folktales probably has far less to do with actual pagan or occult practices than with the fact that old women, the most likely to be suspected of being witches, kept pet cats far more often than they kept pet pigs. And there is some evidence that many surviving folk beliefs about the evil propensity of cats and the dangers of encountering one at night were neither the work of the Church in its campaign of slandering pagan religions nor a reflection of any powers cats were actually held to have among practitioners of those pagan religions, but rather simply the creative inventions of smugglers and counterfeiters—who spread a good story to keep nosy peasants at home at night and away from their illegal traffic and lairs. The fact that cats were so widespread, of course, made them perfect for this role, since anyone going abroad at night was likely to encounter one.

Because cats were ubiquitous, they also lent themselves readily to tales and magical beliefs that enjoyed currency for elemental reasons that had far more to do with the fundamental nature of

The cat was just one of many animals associated with witches. In this woodcut from a 1619 account of an English witch trial, three of the accused are shown with their familiars: a cat, a mouse, an owl, and a dog.

the human mind than with anything specifically about cats or religion per se. There are certain narratives that are simply compelling, that touch a spine-tingling nerve, and the classic ghost story has elements that appear over and over. One of the most ubiquitous and widespread folktales that associate cats with witches—similar stories are found in Scotland, France, Italy, and Germany—is the tale of "repercussion." A man is attacked by a cat. He strikes it a blow that breaks or severs a limb. The next day an old woman in the neighborhood is found in bed with the corresponding limb smashed or amputated. Such an elemental love for the outré and the grotesque is nearly universal in the human psyche.

All that one can reliably conclude from an examination of the many cat superstitions is that the human race is extraordinarily ludicrous and that life in ancient times was sorely lacking in effective medicine or much in the way of entertainment. In Germany it was believed that if a cat jumped onto the bed of an ill person he would die. Elsewhere it was believed that if a cat jumped *off* the bed of an ill person he would die. A cat that left a house would bring bad luck. A cat that *entered* a house could prevent bread from rising and spoil the catch of fishermen. White cats bring bad luck, black cats bring good luck—or the other way around. Cats could suck the breath from a sleeping child. Entombing a cat in the walls of a building being constructed would confer protection. Maiming a cat could shield one from sorcery. Killing a cat would bring misfortune. A girl who treaded on a cat's tail would not get a husband; a man who took good care of cats would get a pretty wife; eating the warm brains of a freshly killed cat could make one invisible; burying a cat alive in a field could prevent weeds from growing; eating cat excrement could cure pneumonia or a bad fall. And there is this recipe for curing an itch ("of any sort"), as reported by Patricia Dale-Green in *The Cult of the Cat:*

> A left-handed man must first find a black cat, and then whirl it three times around his own head. He should then prepare an ointment consisting of nine drops of blood taken from the cat's tail and the charred remains of nine roasted barleycorns. This unguent is applied with a gold wedding-ring as he walks thrice around the patient, invoking the Trinity. If the itch is known to be caused by

shingles, all that is required is to smear over the affected area blood taken from a black cat's tail.

The trouble with reading anything at all into this is that exactly equivalent sorts of magical formulas can be found in which toads, stones, herbs, holy water, amulets, and strips of ribbon feature. Whatever people found around them they used for magic or omens, and the salient features of all these procedures are their complexity, strangeness, and ritual sequences; the choice of the particular thing to be used in the ritual often seems wholly incidental to the object that is to be achieved through its use.

The public rituals in which cats were slaughtered under Church auspices are likewise probably far less indicative of how people or the Church felt about cats than of how superstitious and cruel people were in general during the late Middle Ages and early modern period in Europe. The public torture of animals was a standard feature of life back then, after all, and dogs and hens also featured in such public sacrifices. And, indeed, ancient pagan religions were hardly unfamiliar with the notion of animal (and for that matter human) sacrifice. The practice of killing cats for magical purposes and of throwing cats into fires almost certainly goes back to pre-Christian times. Saint Eloi tried unsuccessfully to ban the practice in the seventh century. Pre-Christian—or at least non-Christian—folk rituals associated with Midsummer Day had long included throwing into a fire objects that were supposed to bring good fortune and avert disaster in the coming year. The practice of lighting of bonfires on St. John's day, which was observed on the summer solstice, was probably an attempt by the Church to co-opt these pagan rites.

So I think it is fair to conclude that the actual position of the cat in human society and the human psyche was far more complex and ambiguous than a simple story of ancient veneration followed by medieval persecution can capture. Cats certainly carried the aura of magical power that, in the historian Robert Darnton's words, is ever "associated with the taboo." Darnton has also convincingly shown that the sexual behavior of cats has loomed large in peoples' ambiguous feelings toward them. But it was ever thus.

That cats may have been viewed as relatively more expendable by the late Middle Ages in Europe, when centuries earlier they were left as treasured legacies in wills, probably has much more to do with simple economics than with religious oppression or repressed sexual insecurities. Certainly by the eighteenth century cats were so common in European cities that they had become a nuisance, filling abandoned buildings, basements, garbage dumps, and public squares, their yowls punctuating the night air. In a society predisposed in no little degree toward drunken violence, the excess cat population was an obvious target. Yet in parts of the world where cats remained scarce they could still be considered extremely valuable and zealously guarded. In 1854, a settler on the Minnesota frontier paid five dollars for a cat at the trading post at Fort Snelling (located in what is now the Minneapolis–St. Paul area), and considered himself lucky to get one at all. After General Custer's expedition to South Dakota discovered gold in the Black Hills in 1874, the nearby town of Deadwood boomed but still had no cats two years later; an enterprising mule skinner by the name of Phatty Thompson saw an opportunity and hauled in a wagonload of cats from Cheyenne, Wyoming, selling them in Deadwood by the pound, with the largest specimens fetching $30—about $500 in today's money. For much of the cat's modern history, scarcity, rather than the myriad magical beliefs that people have held about cats, has determined their fate at the hand of man.

Favorite Colors

Scarcity was definitely a factor in fomenting the one clear genetic divergence that has occurred between the domestic cat and the wildcat, namely color. Because coat characteristics are such an obvious feature of any animal, and because they are controlled by a limited number of genes that usually act in a simple and straightforward fashion, they are one genetic aspect of a domestic species that even fairly witless humans can influence.

For the most part, human beings have never exerted a direct control over the mate choice of domestic cats—not even among the pet cat population, let alone feral cats. Yet geographic evidence clearly points to human selective forces that were at work in producing the rich array of colors that domestic cats come in. This selection was mainly exercised when people moved and had the chance to choose the cats they would take with them to new lands.

The wild-type coloration of the cat is invariably a striped, or "mackerel," tabby, in which a sharply striped dark pattern on the body, legs, and tail is superimposed against a grayish background. The background hairs are not actually a uniform gray, but rather have a sort of salt-and-pepper appearance that is found in many mammals, the result of a band of lighter pigmentation below the tips (this is known as "agouti" coloration, named after the otherwise obscure South American rodent that exemplifies the pattern particularly well).

But among domestic cats, various color mutations appear in a considerable percentage of the population (plate IV). The most important color variants controlled by single genes are the blotched tabby (the *mc*, or t^b, gene), in which the tabby stripes are intensified and often fuse together to form dark whorls or blotches; Abyssinian tabby (T^a), in which the tabby stripes are reduced to a vestigial ticking on the legs; non-agouti (*a*), in which the pigmentation of the hairs is uniform along its length, resulting in a solid-colored cat, typically all black; dilute (*d*), in which the color pigments are scattered in clumps in the hairs, resulting in a smoky blue or fawn or cream color; and orange (*O*). The mutant orange *O* gene has the peculiarity of being carried on the X chromosome, and because females have two X chromosomes while males have only one, this so-called sex-linked trait produces a sexual divergence in color possibilities. A male with an *O* gene is just orange. A female, however, can carry an *O* gene on one of her X chromosomes but not on the other, resulting in the expression of both orange and non-orange coat colors simultaneously; this gives rise to the combined orange and black patterns of the (almost always female) calico- and tortoiseshell-colored cats.

Other mutant genes code for long hair, spots, pure white coloration, and a number of other color variants. A pure white coat in cats also has an odd characteristic that is common to many other mammals. The cells that produce colored skin and hair pigments derive early in fetal development from cells in the neural crest, an embryonic structure that also generates the brain and spinal cord. As a result, you usually can't lose your color without losing your brain as well, or at least parts thereof. Cats that carry the *W* gene, preventing the development of color pigment cells, also often suffer from neural defects such as deafness or blindness.

MAJOR COAT GENES IN CATS

A	agouti (tabby)	*a*	non-agouti (solid color)
Mc	mackerel, or striped, tabby	*mc*	blotched tabby
T^a	Abyssinian, or ticked, tabby	t^a	nonticked
D	dense color	*d*	dilute color
L	short hair	*l*	long hair
O	orange	*o*	non-orange
S	white spotting	*s*	nonspotted
W	pure white	*w*	normal color

Capital letters stand for dominant genes, lower case for recessive genes. A cat must carry two copies of a recessive gene, one inherited from each parent, to exhibit the corresponding recessive trait; a single dominant gene is sufficient to express the dominant trait.

Breeding experiments beginning early in the twentieth century established a direct connection between many color traits and single genes. In the 1940s the renowned biologist J. B. S. Haldane suggested that it would be quite easy to carry out a worldwide research project on the population genetics of cats just by taking a census of cat colors in cities all over the world. His idea was taken up, and over the following three decades scientists from many countries published dozens of papers reporting gene frequencies of cats in various far-flung locales. (At one conference some researchers did note that it wasn't necessarily quite as easy as Haldane had suggested, as in a number of cities biologists prowling about the streets and peering down alleys and up at balconies through binoculars attracted the unwanted attention of the police, who took them for spies or peeping toms.) But by the 1970s the data had begun to reveal some distinctive and fascinating patterns.

Most notably, when the researchers constructed gene frequency maps, they found a smooth but often very steep gradient in the occurrence of the non-agouti, blotched-tabby, and orange genes across Europe, Asia Minor, and North Africa. The steepest gradient was in the blotched-tabby gene. About 65 percent of cats in London are blotched tabby, versus only 4 percent in Iran. Because blotched tabby is a recessive trait, which shows up only if a cat carries two copies of the *mc* gene (one inherited from each parent), these coat-color frequencies correspond to an underlying gene frequency for the *mc* gene of 81 percent in London versus 20 percent in Iran. High frequencies of the blotched-tabby gene also appear in a corridor running from Marseilles to Paris.

The distribution of the non-agouti gene shows some similar characteristics, with a high frequency in the eastern Mediterranean, the Marseilles–Paris corridor, and in England.

A detailed analysis of the maps has allowed scientists to reconstruct the places where, historically, these mutant colors first got a foothold in the domestic cat population and henceforth spread along paths of human migration and commerce. The non-agouti mutation appears to have first arisen in Greece or

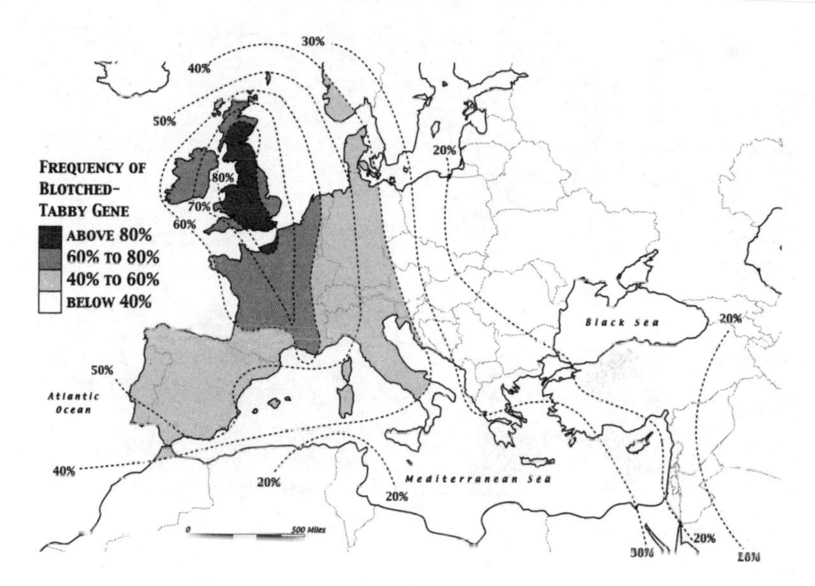

The mutant gene responsible for the blotched-tabby coat color may originally have spread from Iran, across the Mediterranean, and up the Marseilles–Paris corridor. In more recent times this mutation appears to have been selectively favored by the urban habitat.

Phoenicia and moved up into Western Europe along the Rhône and Seine valleys, which have since ancient times been important transportation corridors; that explains the high frequency of the gene along the Marseilles–Paris axis. The blotched-tabby gene may have originally spread from northeast Iran by a similar overland path.

The orange and white genes, however, show a dramatically different distribution. The *O* gene is found across a wide swath in Asia Minor in relatively constant proportions (around 20 to 25 percent) but drops off in almost every direction from there, suggesting that it did not spread beyond Asia Minor by land. Yet *O*

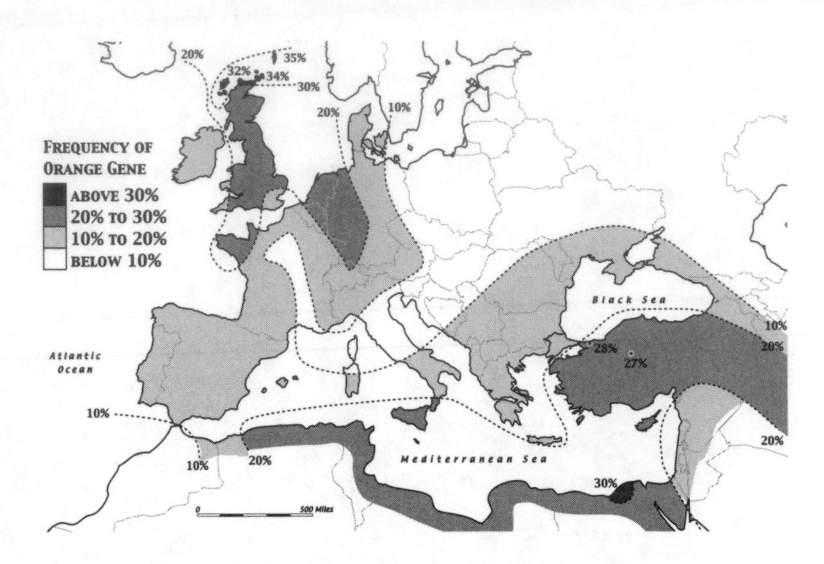

The gene for orange coat color appears to have originated in Asia Minor and may have been spread by the Vikings to the north and west coasts of Scotland and to the Faroe Islands.

also appears at high frequencies in pockets far removed from there—Sicily, the Mediterranean islands of Spain, the north and west coasts of Scotland, the Faroe Islands, and Iceland. The *W* gene occurs at notable frequency together with the *O* gene in some of these same places: eastern Turkey and the North Atlantic islands, most notably. All of this points to transport of orange and white cats by sea. The biologist Neil Todd, who analyzed this data, suggested that the Vikings were probably responsible for this seaborne transport of orange and white cats—probably because they just happened to like orange cats and white cats.

Migration appears to have been a potent force in amplifying the frequency of mutant colors in cats. Whenever people picked some cats out of the local cat population to accompany them on a voyage, they were able to satisfy their fancies for novelty. Todd has noted a curious phenomenon: Many color mutants seem to exist at a frequency that results in a maximum 90 percent/10 percent split between two alternate colorations. The reason is apparently that the value of a novelty begins to decline when it surpasses 10 percent; what is scarce is attractive. The 10 percent rule holds up remarkably well, regardless of whether the trait is inherited as a dominant trait, such as orange, or a recessive trait, such as non-agouti. It even holds for traits such as the taillessness of the Manx cat, which is expressed in cats that carry a single gene for the trait but which is lethal when two genes are inherited.

It also holds for the mutant gene responsible for extra toes, a trait that appears to have arisen in New England in colonial times. Six-plus-toed cats make up about 1 percent of the cat population in Salem and Boston (versus about 0.01 percent in New York). They make up 7 percent of the population of Halifax, Nova Scotia, however—apparently a legacy of the loyalists who fled New England during the Revolution, and who took a disproportionate selection of these novel cats with them.

Complex statistical analyses of cat gene frequency patterns by the Russian scientist Alexander Vinogradov have confirmed that whenever people had a chance to colonize new areas, they took such a "biased colonization sample" with them. The New World has a significantly, and consistently, higher proportion of orange, dilute, white, and longhaired cats than do the Old World areas from which the American settlers came. Across the United States the frequencies of these genes are, however, quite smooth, in sharp contrast to the steep gradients found in the Old World.

This is strong confirmation that the New World–Old World differences are not the result of some subsequent change or evolution that took place in the New World after colonization, but rather reflect what biologists call the "founder effect." A relatively small and "biased" sample was introduced into a wide-open new territory and rapidly expanded, like a gas through a bottleneck. When such an expansion occurs there is very little new selective pressure that can be brought to bear, and the result is a high degree of uniformity across very large geographic areas.

In St. Petersburg (formerly Leningrad), a similar colonization effect has been noted; that city's cat population was completely destroyed, along with many of the city's human inhabitants, during the thirty-month-long Nazi siege in the Second World War. Following the war, a vast human migration to the rebuilt city occurred, providing a local opportunity for recolonization of the cat population as well. The cats of St. Petersburg show a very strong bias with a much higher than average frequency of this same group of "luxury" genes, including the highest frequency of the longhair *l* gene anywhere in the world. Subsequently, the gene frequencies of the St. Petersburg cats has been drifting back toward the more typical profile of Eastern European cats in the surrounding regions.

Adapting to the Urban Environment

These historic patterns of gene flow don't tell the entire story, however, because they do not explain one glaring peculiarity in the gene maps: the frequency of blotched-tabby and non-agouti cats shows a striking correlation to human population density.

The cats of Paris and London share a much closer gene profile than either do with the cats of surrounding suburban and rural areas; consistently, the urban cats are darker than the rural cats. The frequency of both the blotched-tabby gene and non-agouti gene were found to be very significantly correlated with human population and human population density of twenty-eight cities in the United Kingdom and Ireland, ranging from 32 percent blotched tabby in Kirkmichel (pop. 1,000) to 81 percent in London (pop. 9 million). Further data suggest that urban cats on average differ from rural cats at every other gene locus that presents a choice between lighter and darker coats; urban cats have disproportionately high frequencies of nondilute versus dilute genes, non-orange versus orange, nonwhite versus white, and nonspotted versus spotted.

Yet a sociological study of cat color preferences among residents of Glasgow found that this strong bias in the actual, observed coat-color gene frequencies of urban cats was markedly different from the "ideal" cat population that would exist if coat colors corresponded to peoples' expressed choices—which, as Vinogradov's data indicate, strongly favor white, dilute, orange, and longhaired. (Interestingly, expressed preferences were very similar in two very different socioeconomic areas in Glasgow. Also, although about 20 percent of the people surveyed expressed a superstitious belief that certain colored cats were good or bad luck, they were almost exactly divided over whether it was black cats or white cats that had each property. So superstitious belief did not appear to be a factor in determining preference.)

Thus, despite what people say they want in a cat, some force is at work—and at work all the more so in big cities—to keep cats a very different set of colors. Conscious human preferences are actually overpowered.

It is an interesting question how mutant colors came to be established in any cat populations to begin with, even small and local cat populations, because the odds were against it. Mutations in the wild usually die out very quickly. The wild-type, striped tabby color is clearly well adapted for camouflage in a forest environment, and an orange wildcat would no doubt be at a serious disadvantage in attempting to sneak up on prey or avoiding becoming prey itself. Even mutations that are neutral in terms of survival value frequently vanish from a population through the random process known as "genetic drift"; a single individual carrying a mutant gene cannot have much of an impact on the population as a whole. Many of the mutant coat genes that now appear in domestic cats have probably been kicking around in the cat population for eons, since they all can be found expressed in other branches of the felid family: The clouded leopard and marble cat are blotched tabby; the black leopard and black jaguarundi are non-agouti; the cheetah is spotted; lions and pumas are Abyssinian tabby; lynx are longhaired. These same genes probably have existed at a low level as hidden, recessive traits in the wildcat population or arose every now and then as easily generated dominant mutations that were always stamped out in the past. But something about the domestic cat's ecological niche allowed these genes to flourish—and to flourish even without (and sometimes in opposition to) conscious human preference.

In the case of the "luxury" genes, the migration hypothesis provides a good model for how they came to be established. People who moved picked novel cats to take with them. When plunked down in virgin territory, those cats and their novel genes were able to dominate the gene pool and rapidly overcome any countervailing tendencies, either from adverse selection (in the case of genes with lethal or deleterious side effects, such as pure white or Manx) or just plain genetic drift.

But the explosion of blotched tabby and non-agouti must represent some other selective force at work. At first blush it seems even more surprising that this should have happened, because even without intentional breeding, or the opportunity to skew the odds through migration and colonization, there were lots of ways people ought to have been able to tilt the coat-color gene pool in any direction they favored. Pet cats have a much longer lifespan than feral cats, and so simply by choosing to feed and care for select members of the free-roaming cat population people should have been a powerful force in the spread of favored genes. A tomcat who was a color people liked, and who was cared for and fed as a pet but who remained free to come and go, would have had a considerable competitive advantage in spreading his coat-color genes among the cat populace at large.

Lately, however, a paradoxical effect may have countered this. The Glasgow study found that people living in two higher-income, suburban areas of the city were considerably more likely to own cats carrying the white, orange, dilute, and longhaired genes than were residents of two lower-income, tenement blocks. But they were also much more likely to have their cats neutered. In the long run, this process would actually work to selectively eliminate the luxury genes from the city's cat population, by sucking out all the pretty colored cats and ensuring that their genes die with them.

But there is also some substantial evidence that this trend toward darker cats in cities long predates the popularity of spaying and neutering of pet cats. Neil Todd has pointed out that there is a very strong correlation between the percentage of blotched-tabby cats found in English-speaking colonies around the world and the date at which each was settled by English immigrants. The data run from a gene frequency of about 45 percent in New England (settled in 1650) to about 55 percent in the Atlantic provinces of Canada (1750) to 75 percent in Australia (1850). All the data points fall on a remarkably straight line; and if that line is extended up to the late twentieth century it runs right through the present-day blotched-tabby frequency found in major British cities, 81 percent. The conclusion is that urban cats in Britain have been getting darker over the last three centuries, and that each colony—which Vinogradov's "bottleneck" data suggest has experienced little change in cat gene frequencies since they were established—has thus captured a frozen record of the mother country's cat-genetic profile at the time of its original settlement.

There are several theories that have been proposed to explain why darker cats should be favored by the urban habitat. One is that it has provided a sort of protective camouflage, especially as cities became darker and sootier with industrialization. J. M. Clark, the author of the Glasgow study, noted that "cat bashing" remained a popular pastime among the urban proletariat, and presumably darker-colored cats blended into the background better and were more likely to escape notice by the local toughs.

But there may also be a hormonal explanation for the favoring of darker colors, for the genes that determine dark coat colors are coincidentally linked to a number of metabolic pathways that affect size and temperament. Such coincidental connections in

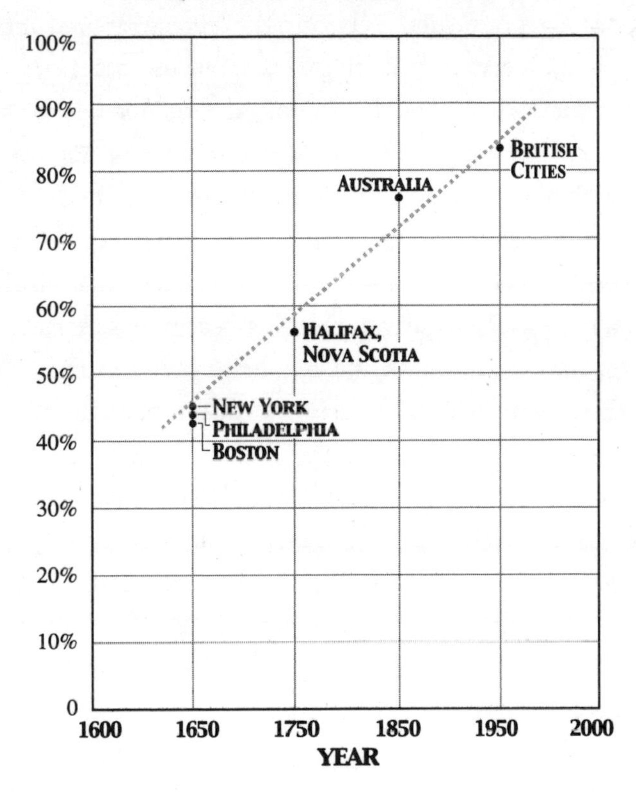

The frequency of the blotched-tabby gene in former British colonies is closely correlated to their date of original settlement. The frequency of this gene has apparently been increasing steadily in the urban areas of Britain that supplied the colonies with their founding cat populations.

the effect of genes is what is known in genetics as a pleiotropic effect; and it has been well established that certain behavioral traits, such as fear and aggressiveness, can be altered simply by crossing animals having certain coat colors. In particular, the skin pigment melanin and the hormone melatonin share a metabolic pathway used to create a chemical precursor that they have in common. Melatonin has powerful effects on sleep, behavior, and mood. It has also been found that among male feral cats, animals carrying the genes for darker coat colors—non-agouti, blotched tabby, and nonspotted—are the smallest and have the lowest adrenal-gland weights. Smaller cats may be more suited to the urban habitat, where there is competition for food resources with many other cats; so, too, cats that possess certain hormonal variations that result in a more placid, tolerant, and less fearful temperament may be better adapted to life in the big city. It may just be a coincidence that certain coat colors come with such advantages.

The Superficiality of Breeds

Only about 3 percent of the owned cats in the United States are registered as a breed; the rest are generally classified as simply "domestic shorthair" or "domestic longhair." There are some thirty-six or so separate breeds of cat recognized by various registries. Yet genetic analysis reveals that the actual differences between them are almost completely a matter of the dozen or so color and coat genes, and nothing more. In other words, what defines a breed is really just its color. Genetically, any given cat breed represents nothing more than the fixing of a particular few of the color and coat genes in a breeding line so that they are passed on consistently to the offspring. Anatomically there are essentially no meaningful differences between breeds, with the one notable exception of the shortened face of the Persian cat.

Breeders often assert that there are distinctive behavioral traits that characterize certain breeds, but hard evidence for this is scant. Rigorous behavioral studies find that behavioral traits vary hugely from individual cat to individual cat in any case, and this individual variation probably overwhelms any average breed-to-breed differences.

It is possible that some behavioral traits just got swept along in the selection for certain color genes. Whenever you genetically isolate one piece of a population, it will, through the luck of the draw, have certain genetic differences from the mean; and the smaller the founding population of a breed the more likely this is. The pleiotropic effects of certain coat color genes may also create genuine behavioral differences in certain breeds. Siamese cats, for example, are often said to be especially vocal. On the other hand, some claims that have been made for the good temperament of certain breeds are unalloyed scientific nonsense. The Ragdoll cat, a breed founded in the 1960s in the United States, was claimed to have a particularly affectionate disposition because the female that founded the line had been frightened in an automobile accident that made her especially attached to her owners. The claim that acquired characteristics can be passed on genetically went out with Lysenko, except apparently among certain cat breeders.

There are several possible reasons that may explain why cat breeds don't really differ very much from one another. It is a curious fact that domestic dog breeds vary in weight by a factor of about a hundred from smallest to largest, while domestic cats span a range of weights of about a factor of two at best. Much of the difference in size and physical shape between various breeds of dog are the result of small changes in the timing and duration of critical periods of growth in the first few months of life. It is possible that cats are, for some subtle reason of their genetic makeup, simply resistant to this sort of genetic alteration. Or it may be that people have never tried very hard to breed cats for any particular type of physical size or shape. Given their potential intractability, it is doubtful that anyone would want a really huge cat. And some of the very marked physical differences in dog breeds may be a largely unintended, pleiotropic effect of selecting for the unique behavioral traits that characterize such specialized breeds as livestock guard dogs, retrievers, scent hounds, herding dogs, and so on. Cats have never been pressed into service in such varied roles, so there may have been little reason for people to exercise the sort of selective pressures that have operated on dogs to produce, if even unintentionally, such an array of physical types. Most cat breeds are probably less than a hundred years old, in terms of constituting a genetically closed population that is bred only among itself. And unlike the kennel clubs that regulate the purebred dog world, the cat breed organizations often permit outcrossing, which helps to maintain genetic diversity, especially in breeds with small numbers of registrations. (For example, the Cat Fanciers' Association permits American Curls to be outcrossed with domestic longhairs and shorthairs, Somalis to be outcrossed with Abyssinians, and Devon Rex to be outcrossed with British Shorthairs.)

In any case, the genetic evidence clearly shows that cat breeds have never been subjected to any very intensive artificial selection, as purebred dogs and Thoroughbred horses have. The geneticist Stephen O'Brien, who has been leading the U.S. National Cancer Institute's efforts to map the cat genome, notes that domestic cats remain extremely outbred. If you cross two closely related cats, such as brother and sister, ninety-nine times out of a hundred their offspring will suffer from a lethal defect or will be sterile, causing the line to die out immediately. That is a strong indicator that there has been extraordinarily little inbreeding in the cat's recent genetic past: Lines in which inbreeding had been going on would have largely weeded out such lethal or extremely deleterious recessive traits, and only those offspring who did not inherit those "bad" genes would survive. (Inbreeding, however, does tend to bring out and fix in the breeding line harmful recessive traits that are not completely or immediately lethal. That is why many inbred lines suffer from lower viability of offspring, lower fertility, and a high incidence of inherited disorders such as the hip dysplasia, cancers, and metabolic diseases that afflict many lines of purebred dogs.) In an outbred population, lethal recessives tend to accumulate in the gene pool but conversely cause few problems, since they are harmful only when two carriers of the same defective gene mate—and it is only closely related individuals who tend to carry the same defective genes.

A final piece of evidence comes from studies of DNA microsatellites. These are strands of "junk" DNA that exist throughout the genome and which are the basis of many DNA fingerprinting techniques. "Junk" DNA is junk because it probably does not actually do anything. It just accumulates from generation to generation; and precisely because it has no active function, mutations can occur frequently in these sequences with no harmful effects, and so are not weeded out through natural selection. The result is that a given microsatellite often will come in many different alternative variants that differ from individual to individual. Marilyn Menotti-Raymond, a researcher at the National Cancer

Institute, analyzed cat microsatellites when the Royal Canadian Mounted Police asked her if it was possible to use DNA fingerprinting to match a cat hair found at a crime scene with a cat owned by the suspect in the case—a grisly murder of a thirty-two-year-old woman who disappeared on Prince Edward Island in 1994. (A match was found, and the defendant, the woman's estranged husband, was convicted.) Menotti-Raymond has since received funding from the U.S. Department of Justice to build a genetic database covering the various breeds of domestic cats.

One spinoff of this research is a comparison of the microsatellite profiles of purebred versus random-bred cats, which tells something about how inbred each is and how much each differs from the others. An individual from a highly inbred line will tend to inherit the same microsatellite variant from both parents at a given spot in the genome (this is known as a "homozygous" state), while an individual from an outbred line will tend to have two different microsatellite variants at any given location ("heterozygous"). Surprisingly, a comparison of random-bred cats, Persians, Abyssinians, and Manx cats by researchers at the company PE Zoogen found significant differences in heterozygosity among these different breeds at only a few microsatellite locations. One might naturally expect the random-bred cats to be more outbred, and thus more heterozygous; and at a few microsatellite locations they were (for example, at one site only 30 percent of the Persians but 70 percent of the random-bred cats were heterozygous). But at other sites the purebred cats were indistinguishable from the random-bred cats, and at still other sites the purebreds were actually *more* heterozygous than the random-bred cats. Overall, it was pretty much a wash. In a preliminary survey of blood samples drawn from twenty-nine different cat breeds, Menotti-Raymond found much the same pattern: No single breed stood out as being more homozygous, and thus more inbred, than any others. She also found the overall levels of heterozygosity to be extraordinarily high in domestic cats, with figures of 90 percent not uncommon. That degree of heterozygosity is much greater than in purebred dogs and Thoroughbred horses, and on a par with mice and human beings, who are well known for mixing it up when it comes to mate choice.

Unlike pigs, horses, chickens, dogs, goats, cows, and sheep, the differences between cat breeds are little more than skin deep.

THREE

The War Between the Sexes and Other Oddities of Feline Society

If I were to impanel a group of scholars—say, a psychologist, a neuroendocrinologist, a cultural historian, an anthropologist, and an evolutionary biologist—and ask them why, exactly, my teenage daughter listens to very bad music, I would likely get a very different answer from each. The psychologist would no doubt talk about peer pressure and developing self-identity; the neuroendocrinologist would discuss the role of hormones in the adolescent brain; the cultural historian might point out the social forces that caused musical tastes to change over the course of the last generation; the anthropologist would draw analogies to tribal rites of passage; and the evolutionary biologist might suggest that rebellion against parents upon reaching sexual maturity is an adaptive behavior that evolved as a way of ensuring that offspring disperse over a wide geographic area, maximizing their reproductive fitness.

All would probably have gotten hold of a piece of the truth. Anything as complex as the behavior of people is hard to explain fully at a single level of analysis. Evolution, history, individual experience, personality, and group dynamics all play a part; and each of these forces acts at a different level, from molecules to civilizations.

The behavior of the cat is the product of millions of years of evolutionary adaptation to a particular way of life; but like people, cats today live in a world very different from that which their wild forebears evolved in. They are the product of history, cultural tradition, and individual psychology as much as evolutionary destiny. The range of social organizations that domestic cats tolerate, and even create on their own, is extraordinary testimony to the complexity of forces that shape contemporary feline behavior. Animal behaviorists used to blithely categorize the domestic cat as "asocial," just like its wild counterpart, but studies in recent years unmistakably reveal that to be far too simple. Free-living domestic cats have been found in groups ranging in size from one to fifty-two cats; individual cats have been observed to have home ranges that can vary in size over a factor of ten thousand, from 0.03 to 760 hectares (about 0.1 acre to nearly 2,000 acres); and the population density of cats in town and country has been noted in various studies to vary from about 1 per square kilometer (about 2.5 per square mile) to two thousand times as many.

It might seem that given such striking differences between the highly variable and (often) highly social behavior of domestic cats, and the uniformly solitary nature of wildcats, there is little point at all in looking to the cat's evolutionary ancestors if one is trying to understand what makes him tick. Yet however much we do need to call upon the forces of recent history and of human and feline culture to explain what makes domestic cats the creatures they are, they still are creatures of their evolutionary past. The things *Felis silvestris* does and does not do in its four-thousand-year-old role as domestic cat are things that it must always have had the biological potential to do, given how scant the biological differences are between the domestic and wild populations.

So the roots of the often contradictory social instincts of the domestic cat must be there to find in the evolutionary forces that shaped the species. The key turns out to be looking for those roots in the right place. For unlike dogs and horses, which transferred to the domestic world much of their social instincts intact from the social structures that existed, and that had an equivalent purpose, in the wild, the social behaviors of the cat largely slipped in through the back door. They were behaviors that existed for other purposes in the wildcat and which—under the historical influence of man and the amazing powers of cats to pass on "cultural" tradition once acquired—have been recycled, retreaded, and given a new lease on life in the domestic cat.

At the same time, though, the domestic cat has retained many of the asocial, and even antisocial, behaviors of the wildcat: its territoriality, suspicion, hostility to novelty, and propensity to acquire such unwelcome habits as spraying urine and fighting with newcomers. Sometimes the social and antisocial instincts of the cat exist in amazing juxtaposition, as in the not uncommon phenomenon of the cat who purrs and rubs against his owner begging to be petted, only to sink his claws into the hapless human and dart away a minute later.

We Want to Be Alone

As sociable as many domestic cats are, they are heirs to some powerful evolutionary forces that continually tug the other way.

There are some basic principles in nature that determine whether the members of a species live by themselves or in groups. Not surprisingly, the most basic of these are food and sex. Mammals that live in rich forests where food is reasonably abundant and more or less evenly distributed tend to space themselves out. The females defend exclusive territories big enough to supply their food requirements. A male in search of a mate is then pretty much stuck with a single female, since the females are too far apart for him to claim and guard more than one at a time. While males may fight over control of territory, once the territorial lines are drawn that settles the matter; mating is determined strictly by who occupies which patch of ground. Mating between males and females may be monogamous, but it is often weak and fleeting; sex is determined by property rights rather than social bonds.

Where resources are more abundant, but still widely and evenly dispersed, the females' territories may be closer together, allowing one male to cover the territories of several females, which he can still successfully patrol against intrusion by any rivals. Because that means the range of several males can also overlap, contests between males for mates tend to be more intense, but still the social tie between males and females remains minimal.

Groups or packs, usually controlled by a single dominant male or pair, on the other hand generally arise only when food resources are patchy and shifting. An individual can no longer satisfy his food needs by staking out a patch of defensible fixed ground and staying put. Animals that lead a nomadic life in search of food may gain several advantages by banding together, such as mutual defense against predators and cooperative hunting or foraging. But mainly the social group is simply a reproductive necessity for a wandering species; lacking fixed territories that define which females belong to which males, the social glue of a herd or pack is the only thing that allows a male to claim and hold a female. Group-dwelling animals usually develop lasting and stable social ties with other members of the group that, while at root about sex, tend to transcend immediate reproductive needs or impulses. These social ties permit sex to be freed from the chain that ties it to a fixed patch of ground in more solitary species. But the instinct to bond with other members of the species is expressed in many ways in group-dwelling animals. Members of a pack or herd know and (often) simply like the other members of the group, and "friendship" bonds between same-sex individuals are a frequent feature of herds and packs. Most grazing animals, like horses, cows, and sheep, fit this pattern of social structure; so do wolves, in which cooperative hunting is often a necessity because their prey is not only shifting and patchy but also is often very large. In their relationship with man, these animals clearly accept human beings as honorary members of the group, and they call upon their basic repertoire of group social skills in their dealings with man.

Cats on the other hand fall squarely in the middle pattern, in which females defend solitary territories and males range over the territory of several females.

There is much that follows inevitably from this evolutionary fact. One is that cats start with a strong propensity to be homebodies. Dogs are happy to go anywhere as long as they are with the familiar members of their group. Cats like to stay home.

There are certain habits of domestic cats that are ineluctably solitary. No matter how sociable and friendly they may be toward other cats and people, cats always hunt by themselves. And like all animals that evolved to be solitary and territorial, cats have strongly instinctive methods of advertising their territory. Fighting is costly for both winners and losers, and so it usually pays to avoid a fight. As a result, these advertising schemes are often highly elaborate and designed to allow occupants of adjacent territories to stay out of each other's way. Birds that announce their territory by broadcasting songs have a remarkable ability to judge the distance, and sometimes even the age and size, of a potential rival through the tonal characteristics and patterns of his song. Likewise, the urine and feces of territorial animals often can be read by other members of the species in remarkable detail. Odors convey information about the individual identity, sex, dominance status, and estrous status of the animal that left them; and apparently in some cases can even indicate how recently an animal passed through the area.

People often think of animals' territories as the equivalent of a chain-link fence demarcating a suburban yard; and it is true that the urine marks left by territorial animals often tend to be more frequently found along the boundaries of their range. Likewise, birds that announce their territories with songs tend to focus their vocalizing along the edges of their territories. But in fact most territorial animals that use urine and feces to establish their territories mark throughout their range; boundaries just happen to be the places where they are most likely to encounter a stranger or a stranger's scent marks, which often trigger their own scent-posting responses.

Cats, both male and female, spray urine frequently on prominent vertical objects throughout their daily travels. Actual observations of the marking behavior of free-ranging cats (this is what graduate students are for) noted that females sprayed on average once an hour, toms a dozen times an hour. Males often mark an object every 5 meters or so along a path. When spraying, cats raise their tails at a 45 to 90 degree angle and aim the urine in a fine spray, typically at a tree, post, or other upright object. This posture is markedly different from the one cats use when simply emptying their bladders; then they just squat. Also it appears that when spraying, cats release a distinctive scent into the urine stream that is not present otherwise. Experiments have shown that cats will spend more than twice as much time sniffing sprayed urine versus ordinary cat urine.

The particularly characteristic smell of the sprayed urine of tomcats has been identified as the product of an amino acid appropriately named felinine, a sulfur-containing compound. Members of the cat family are the only mammals that excrete this compound in their urine. Like territoriality in general, the spraying of felinine seems to be triggered by testosterone; intact tomcats excrete about three times as much as castrated males, and about five times as much as females. Chemists who have artificially synthesized felinine in the lab have found that the pure substance actually has no smell, but after being stored for a while it develops a noticeably "catty" odor, apparently as it breaks down into some related compounds. This delayed reaction may enhance its value as a territorial marker, for it serves to keep the smell of a sprayed-urine mark alive longer.

Feces also play an important signaling role in cat territoriality. Studies of free-ranging cats have found that, contrary to popular belief, cats most definitely do not always cover their feces, or "scats." They do so only about half the time, in fact. Cats are much more likely to cover their scats when close to the core of their territory, however, especially the areas that include their habitual resting and sleeping spots. Outside the core area scats are frequently left uncovered. The conclusion that scats serve a signaling function is reinforced by the fact that when in their usual sleeping or resting spots, cats usually move a few meters away before defecating, but outside their core range deposit their feces right on the trail where they are traveling. This same pattern of covering feces close to home but leaving them in the open when abroad has been observed in wildcats. Covering scats close to home may serve a useful purpose in preventing the spread of parasites (cats notably do not like to eat in the same area in which they defecate) and also help the cats to avoid broadcasting the location of their home to potential predators.

The instincts of territoriality and marking territory are ever present in domestic cats. They may be modified through social circumstances, learning, or more direct manipulation such as drugs or surgery that alter hormone levels, but they are a basic part of the domestic cat's heritage that, indeed, constitute some of the strongest instincts that exist in nature.

The War Between the Feline Sexes

One particularly nasty consequence of the cat's solitary evolutionary heritage is in the relation between the sexes. The territorial and social structure of the cat is just about ideally optimized to encourage violent fighting between males. There are no permanent matings between males and females; the range of each male overlaps with several females and thus, potentially, of several other males; and females have a definite breeding season, with the main period of sexual activity occurring in late winter–early spring and a second, less active period in early summer. All of these factors ensure that males will face intense competition for access to mates. In species like the cat in which females occupy fixed territories and males overlap with several females' ranges, there is usually a marked sexual dimorphism: Males are bigger and have specialized equipment for fighting with other males, such as horns or antlers. Male cats are indeed larger and considerably more aggressive than females, and they have larger jowls and an extremely tough layer of skin guarding the neck, an adaptation to the demands of fighting.

The tendency toward fighting between males is almost exclusively determined by the male hormone testosterone. When male cats are castrated in adulthood, about 90 percent stop fighting altogether, 50 percent immediately and 40 percent after several weeks or months of gradual tapering off. (Likewise, about 90 percent of male cats castrated before puberty never engage in serious fighting with other males.)

The strategy of fighting one's way to reproductive success is actually remarkably ill suited to the modern environment that most domestic cats occupy, but it persists nonetheless. In the widely dispersed wildcat population, a male who seeks to claim exclusive right to the territory of several females will typically need to fight off only one intruder at a time. Aggressively meeting any intrusion is thus an effective strategy. The most aggressive male in such circumstances generally ends up being the one who gets to pass his genes on to the next generation. But most free-roaming pet cats and feral cats live in much higher concentrations, and many males may crowd around a single female in heat—as many as six males at a time may court a given female. Studies by Eugenia Natoli of a large stray cat colony in central Rome found that under such circumstances it is often the less dominant and aggressive males that are actually more successful in mating with females. What seems to happen is that while two of the meanest and toughest males are engaging in a prolonged and violent duel, one of the less aggressive males takes advantage of their distraction and slips in and mates with the female. The aggression of male domestic cats might thus, in Natoli's words, be "a suboptimal remnant of the original behavior" of wildcats. If so, it is another good example of how domestic cats remain creatures of their evolutionary heritage, despite their very changed social circumstances.

Mammals that follow the wildcat's solitary social structure generally have mating systems in which the relationship *between* the sexes is highly competitive as well, and sometimes downright antagonistic. Because solitary females cannot count on finding a mate whenever they are in heat, they are usually induced ovulators: That is, they do not ovulate spontaneously according to an automatic schedule in their estrous cycle but rather release eggs only after copulating. Studies in cats have established that it is the stimulation of vaginal nerves during copulation that sends the requisite signal to trigger ovulation.

As a result, evolution has smiled upon male cats who push the limits of vaginal stimulation just as far as they can and still manage to copulate. The penis of the male cat is covered with small spiny projections that—only after intromission and ejaculation—cause a very strong, and apparently even painful, stimulation of the vagina upon withdrawal. This appears to be the reason for the remarkable postcopulatory behavior of female cats, which has astonished human observers for centuries: The female routinely turns around and violently cuffs the male on the nose immediately after he withdraws.

Male cats have occasionally been observed to engage in an even more violently competitive reproductive strategy: killing kittens sired by a different male. When a female's kittens die, she quickly stops lactating and goes into estrus again. Thus a male who dispatches his rival's offspring can sire a new litter of kittens sooner. Infanticide in domestic cats is rare, however, and Natoli argues that that is probably because the advantage is slight. In contrast to lions, in which infanticide by males is a well-established phenomenon, domestic cats have a short gestation period (about 65 days versus 110 days) and a much shorter nursing period before weaning (60 days versus 9 months). Also, domestic cats have highly seasonal mating periods, whereas lions can breed all year round. Thus a male cat probably doesn't gain that much time by knocking off his rival's offspring.

Female cats for their part have evolved a number of highly sexually antagonistic reproductive strategies of their own. Many of the behaviors of females in heat seem designed to coyly induce competition among males. Females in heat call and increase their scent-marking activity notably, communicating their availability far and wide. During the first few days they are in heat, and while they are doing all of this advertising, they are not receptive, and thus will not mate with whoever happens to be the first to come along. When surrounded by several courting males, a female will sometimes suddenly dash from one spot to another, forcing the males to scramble to reclaim a favorable position close to her. All of this helps to ensure—at least in the ancestral wildcat social structure—that it is the healthiest and fittest male in the area who succeeds in mating with her.

The competition does not end there, for female cats are highly promiscuous and typically mate with several males if given the opportunity. That allows for sperm competition to occur within her genital tract; basically it is the healthiest and fastest sperm that reaches and fertilizes an egg.

The female strategy of mating with several males may also be a way of maximizing genetic diversity among offspring, if, as appears likely, cats are capable of bearing a litter in which different littermates have been conceived by the sperm of different males. Finally, multiple mating is an effective female counterstrategy for foiling the male strategy of infanticide; it sufficiently muddles the question of paternity that a male cannot be certain which kittens are his.

Reengineering the Social Tool Kit

Fighting, territoriality, urine spraying, hostility to and suspicion of strangers, and an inability to adapt to new surroundings are behaviors that no cat owner would deny can be found, at least to some degree, in domestic cats. All of these behaviors serve clear evolutionary purposes in the solitary wildcat, and whether or not they continue to serve a useful purpose in the domestic cat, they remain. A few thousand years of domestication, under which extremely little artificial selection by human beings has been brought to bear, is no match for millions of years of evolution and intense natural selection. Understanding the evolutionary origins of these behaviors, and the elemental motivations that lie behind them, is a key to trying to do something about them, a topic I will return to in Chapter 7 in the sections on problematic and abnormal behaviors in pet cats.

But cats also exhibit sociable instincts that are not at all part of the inherited repertoire of social norms that govern cat-to-cat relations among wild adults.

Where do these behaviors come from? Cats, as I have said, have a basic territorial instinct to bond to places more than to other beings. Yet every cat owner knows of cats that are extremely companionable. One of my barn cats, Shawn, will go to great lengths to be with people. He will follow members of the family on walks down the road and into the woods, for a mile or more (for years Shawn was a fixture of my children's Halloween trick-or-treat outings); he will even regularly attempt to sneak into the house by insinuating himself beneath a passing dog.

Following people around is admittedly unusual even among pet cats. But extremely common among all domestic cats are such social interactions as playing with, rubbing against, grooming, and lying close to other familiar cats and people. Very careful quantitative studies of free-living barn cats have found that while the cats did not ever organize their movements together—the probability that any two given cats would be in the same place at the same time was never greater than what pure chance would dictate—the cats did spend a lot more time very close to one another than expected from a purely random distribution. In other words, when they happened to be in the same place at the same time, they actively sought contact with one another. Whenever an adult cat was in the barn, for example, it was within a meter of another adult cat about 20 percent of the time. When asleep, adult farm cats were observed to be in direct bodily contact with one or more cats about 50 percent of the time, even in very hot weather. Each cat showed definite preferences for which other cats he spent time with, groomed, and slept next to. These friendly behaviors occurred particularly between females and other females, and between males and females, but also were common between neutered males and other neutered males.

So, again, where does a solitary species get off acting this way? Animal behaviorists sometimes want to try to explain every last thing an animal does in terms of evolutionary adaptation, the argument being that the only reason a behavior could exist is if it translates directly into increased survival. This approach unquestionably works well in explaining the main things that animals do. Wasted activity is usually costly, if not downright dangerous or fatal, and so it gets weeded out over the course of evolution; what is left are the things that really do play a crucial part in helping an animal survive and pass on his genes.

But there is a danger in being too insistent about this, because there are clearly some things animals do that they just do, which are not particularly adaptive at all. Sometimes they are the excess spillover of adaptive behaviors; sometimes they are a novel concatenation of two or more adaptive behaviors; and sometimes they are simply what happens when an animal is trying to get from adaptive behavior A to adaptive behavior B and the only way to do that is to pass through behavior C, which may be odd, absurd, pointless, or comical. Many pregnant dogs or other mammals, for example, if given a stuffed animal, will pick it up and carry it around like an infant and even lie down with it and try to make it nurse. There is no conceivable adaptive purpose that explains this particular behavior; it is just that over the course of evolution, mothers who have whatever genes confer especially strong maternal instincts are more likely on average to be successful at raising offspring, who in turn will carry these genes and pass them on to *their* offspring. The wasted time and energy involved in occasionally carrying around stuffed animals is a small price to pay for the offsetting advantages that follow from possessing an overabundance of motherly instincts, especially since there are relatively few stuffed animals in the wild. (Occasionally the overflow of maternal instincts will lead a mother to "kidnap" the offspring of another female; this happens once in a while in a farm flock of sheep, for example; and a ewe with a single lamb of her own may end up nursing both her own lamb and a stolen lamb. This behavior surely would over the long run lead a ewe to be less successful in passing on her own genes. But as long as it doesn't happen too often, natural selection will still favor a very strong maternal instinct, given the very great advantages it confers on the ewe's own true offspring, in being protected against predators and receiving adequate nutrition.)

The social behaviors of adult domestic cats likewise appear to be largely made up of behavioral leftovers, overflows, and hodgepodges. There are several times in the natural life cycle of the wildcat during which cats do have good reasons to interact with others of their kind: mating, in which males and females briefly come together; the raising of young, in which the mother performs a variety of important social roles, most notably in teaching kittens to hunt; and infancy, in which kittens live as a group and interact in a variety of ways with their littermates, most notably in play, which is a distinctly youthful trait in many mammals. All these behaviors seem to have been cannibalized by domestic cats to create their adult social behaviors.

Uses of Play

Play is itself a behavioral oddity, which may be less an adaptive or purposeful activity than just an evolutionary coincidence. Animal behaviorists have long viewed play as something desperately in need of an explanation, because it appeared to be very costly in terms of energy expenditure for the animals that engage in it. So various theories have been proposed to explain what makes play pay: perhaps it is a way for growing animals to get needed exercise; perhaps it is a way for them to learn important adult social skills; perhaps it is a way for them to refine the motor coordination required for hunting.

But some more recent studies, in cats and in other species, raise the distinct possibility that play is nothing but an accident of the way many animals develop from infancy to adulthood. Infants do not need to chase prey, fight with sexual rivals, defend territory, or mate. Adults do. These behaviors thus need to develop over the course of the transition from infancy to adulthood, just as many physical traits do. Acne serves no adaptive purpose in human teenagers that anyone has identified; it is a byproduct of the bodily changes that take place in the transition from infancy to maturity. Likewise adult behaviors as they develop have to phase in; it is hard to conceive of a way that a kitten can one day be acting like a kitten and the next day like a full-fledged adult. Play may thus be a manifestation of the inevitable confusion and disarray that occur as these behaviors begin to appear. Indeed, most play is made up of seemingly incoherent bits and pieces of adult hunting, escape, and sexual behavior. In kittens it typically involves pouncing, chasing, stalking, rearing up, wrestling, and mounting.

Some of the best evidence that play is not in fact very costly in terms of energy expenditure comes from studies of kittens. Ten- to twelve-week-old kittens spend about 9 percent of their time playing. By carefully measuring the amount of oxygen consumed by kittens while playing within a test chamber in the lab (the air being pumped in and out of the chamber was monitored by an oxygen analyzer), Paul Martin of Cambridge University calculated that play accounted for only 4 percent of the kittens' net daily energy expenditures. Even more interesting is the discovery that one reason kittens play a lot but adult cats don't is that kittens simply have more playmates and more opportunities to play. Social play in cats declines in part because kittens begin to show fully adult behaviors as they approach sexual maturity, which occurs at around 5½ months. Beginning at around 4½ months males

Stalk

Rear

Wrestle

Common social-play postures in cats: stalking, rearing, and wrestling

begin to mount and grasp females by the neck, and females start to avoid them as a result. Even more important, starting at about 12 weeks kittens begin to spend an increasing amount of time away from the nest. Kittens who have free access to the outdoors spent about 2 hours a day outside at age 12 weeks but 4½ hours by age 16 weeks. From 12 to 19 weeks the frequency of play declines dramatically as well; so does the frequency of other social behaviors between littermates such as mutual grooming. By 4 months, kittens in free-living conditions are generally fully weaned and have to find their own food, and so they simply have much less opportunity to spend time with one another.

The very fact that play really does not serve any very sharply focused purpose has probably permitted it to elude evolutionary pressures that might otherwise have worked to stamp it out completely by the time an animal reaches adulthood. All adult cats retain some propensity for play; most adult cats continue to play by themselves, pouncing on imaginary objects or assuming a play-fighting stance by themselves. What they chiefly lack, in their ancestral wild condition, is the opportunity to exercise these proclivities in play with others.

In domestic cats, however, such opportunities are often plentiful—among colonies of barn cats or feral cats, and between house cats and their human owners. Play occurs even more often in these domestic settings because the people and other cats often initiate and actively encourage play themselves. Not that that is strictly necessary. I once had a cat, Meatloaf, who even saw in the reluctant dogs of the household an opportunity for play. Meatloaf was extremely adept at lying in wait and then ambushing a passing dog, springing out and swatting the dog on his nose before he knew what happened. I can't say that the dogs viewed this as much of a game, but they put up with it, and that was enough to encourage Meatloaf to persist.

Behaviors associated with maternal care and mating are more certainly adaptive instincts than play may be, but they are also exceptionally strong instincts and as such are particularly prone to bubbling over into new situations. Grooming is an essential part of the physical care and bonding between mother and kittens. Until kittens reach about 5 or 6 weeks of age, they depend upon their mother to lick their urogenital region to stimulate urination. The point is that the instinct to lick another cat is part of the social tool kit that evolution has equipped mothers with; but as with most maternal instincts it is not something that is easy to arrange in the mammalian psyche to be precisely fine-tuned or turned on and off with exact precision. (Humans certainly exhibit a good deal of bubbling over of maternal instincts, too; most people find small things with big eyes and big heads cute, whether they are human babies, puppies, pandas, or even cartoon characters.)

Food begging in kittens and food sharing by mothers is another such instinct. After all, the maternal instinct to share prey she has caught has to be an extremely powerful one—powerful enough to overcome the competing instinct to eat it herself. Starting when her kittens are a few weeks old, a mother cat will begin to bring killed rodents back to the nest; when the kittens are about 4 weeks old she will bring live rodents and release them in front of the kittens and allow the kittens to try to seize the prey themselves. If they fail, she will sometimes recapture the rodent and release it again.

This basic instinct may explain why adult cats, even when by themselves, frequently play with a live animal they have caught, rather than neatly dispatching and consuming it. Like grooming and play behavior in general, this instinct is likely to be activated particularly, however, when adult cats are permitted or forced by circumstances to live in a social grouping. Many cat owners have experienced being on the receiving end of this behavior, finding dead rodents, birds, and reptiles placed at the back door. In one sense the cat is treating his owner as an honorary kitten; in another sense the cat is just doing something that cats tend to do around other cats, given this strong innate instinct. In wildcats, the only opportunity that normally occurs to trigger the instinct is the presence of a litter of kittens. In the domestic cat, the opportunities are frequent and varied; and adult domestic cats living in social groups on farms have even been observed to share prey with other adult cats.

Cats that share a house or a farm will likewise also often groom one another, and pet cats regularly "groom" people, too, licking them and seeking close physical contact.

The Adaptability of Loners

Individual cats do differ markedly in their readiness to exhibit these social instincts toward others in adulthood, and, as I mentioned earlier, one important factor is the experience they themselves had as kittens. That there should have existed for millennia a large pool of free-living village and farm cats that spontaneously show a tendency toward sociability as adults probably reflects a cultural "tradition" that has been perpetuated by the cats themselves; kittens that grow up interacting with strange adults are much more likely to continue to engage in such social interactions themselves as adults. The ways that human owners can influence a cat's future sociability by deliberately manipulating the conditions of his kittenhood is a subject I will return to in Chapter 6.

But the striking thing about cats is that none of this is set in stone. Paul Leyhausen, a pioneering researcher on the social behavior of cats, concluded that there was nothing automatic about the social structure of cats. A dog is usually miserable by himself. A cat, even a cat brought up in a social group, can still do just fine on her own. What determines whether a cat society is cohesive and stable or loose and individualistic is ultimately a complex interplay of individual characteristics and opportunity. When food resources are concentrated and abundant—as on dairy farms or in urban areas where cats are fed handouts or where garbage dumps provide a cornucopia—females generally form large, stable groups that persist for generation after generation.

In other words, when brought together coincidentally by the availability of copious food—or when brought together deliberately by people who keep multiple cats in a household—cats have an opportunity to express whatever latent potential for sociability they have. The chance to express these sociable behaviors results in real and lasting ties, but they are ties of a very different sort from those found in innately and necessarily social animals like dogs. Dogs who live in a household seek in effect to re-create the wolf-pack structure with the other members of the household, human and canine. A pack is like a conventional army unit, with hierarchical ranks from four-star general to buck private. Everyone has his place to fit into, and a highly developed system of establishing and signaling dominance and submission ensures that each member of the group finds his place, knows his place, and usually sticks to his place. The social hierarchy is the framework that wolf society is built upon, and it is based on leadership and acquiescence to that leadership.

But cats who find themselves living in a group are more like commandos who have been trained to carry out daring solo missions on their own, and now find themselves with a bunch of other commandos all on exactly the same mission. Although each one pretty much keeps doing what he's trained to do, it's only natural that along the way some friendships will spring up. Although cats do express dominance and submission, especially in aggressive encounters between males, the normal outcome of a dominance encounter is that one cat slinks off and cedes territorial control to the other. When cats are artificially forced into overcrowded situations or into groupings with strangers, they will engage in dominance contests. But what emerges is not a stable, linear dominance hierarchy as seen in wolves or horses but usually a "despotic" hierarchy in which a single top-ranking individual is deferred to by all others in the group. Sometimes a cat emerges on the bottom of the totem pole as well, and becomes a pariah who is picked on by all others. A pariah cat may spend most of his time hiding or crouching in a corner if he is physically restricted from doing what he would really like to do, namely get the hell out of there. Often any movement at all by a pariah triggers an attack from one of the other cats.

But even the top cat does not exert the sort of group control commonly seen in social animals. The cat who is the most successful in dominating others for access to food is often not the most successful in commanding the greatest freedom of movement for himself or controlling the movement of others, which are normal measures of social dominance. Nor, experiments found, was the dominant cat always the one who turned out to be the most bold in investigating or approaching novel stimuli, like a remote-controlled car, an electric fan with streamers, a vacuum cleaner, or a parrot in a cage. Again, this is very different from the pattern seen in wolves and other social animals that automatically form groups.

So hierarchical dominance, so consistent a feature in obligatorily social animals, is not the foundation of cat society.

Rather, the house cats, farm cats, and stray and feral cats that live amicably in groups each tend to keep doing their own individualistic thing, but the difference is that they now do it in more or less exactly the same place as another cat whose presence they tolerate—and who is also doing his own thing. Pet cats living in the same household who are allowed outdoors have home ranges that overlap more than 50 percent with one another. But their ranges remain mostly separate from the ranges of cats from other households. Likewise, free-living cats studied in a Japanese fishing village formed two very large groups, each centered on one of the two dumps that were provisioned each day with fish scraps. The cats in each group had ranges that heavily overlapped with other members of their group but not at all with members of the other group. Cats living in the Portsmouth dockyards in England showed the same pattern, forming twenty-eight mutually exclusive groups, each centered on their own garbage bin.

The superposition of a bunch of cats each exhibiting its individual territorial behavior toward strange cats, and each occupying a territory that mostly overlaps with other members of the accepted group, does have the net effect of creating exclusive group territories. The individual territory of each member usually overlaps completely in the "core area," such as the house or barn, where the cats rear their young or spend most of their time. Members of the group each repel strange intruders, male and female, from their own home range; and the net effect is that strangers to the group are kept out of the whole group's area, and the group's core area in particular. But this, too, is very different from what happens in wolf society: A wolf pack jointly owns a territory and all members of the group move freely throughout it, whereas the territory of a cat social group is really the sum of the individual territories of each member.

One of the most intriguing features of such cat societies is that the females frequently rear their young communally. Kittens of different mothers may share the same nest and may be nursed by females other than their own mother, too. This is a consequence of overflowing maternal instincts and the opportunity that proximity to other females provides for that overflow to fall on the kittens of another female. It may also confer some selective advantage, though there is a great debate in the scientific literature on this point. It is possible that the banding together of females in this fashion affords safety in numbers against infanticidal males, for example. This may be particularly important during the frequent moves of nest site that the mothers make during the first few weeks after the kittens are born. Shifting nests may itself be a way to protect against predators or infanticidal males, but a solitary female has to leave the rest of the litter alone repeatedly as she carries each kitten by the nape of the neck to the new nest.

Kittens in communal nests, however, are rarely if ever left unattended, even during moves, when they are usually shuttled in a relay by the several mothers.

It is also possible that because many of the mothers who share duties in raising kittens this way are close relatives—often mothers and their adult daughters—a female who nurses another cat's kittens in this communal system is still helping to ensure the survival of at least some of her genes, and so gets a payoff for her efforts that way.

A final theory is that by being nursed by different mothers, kittens acquire a broader range of maternal antibodies and so are more resistant to disease. On the other hand, there is at least some evidence that communally reared cats are more susceptible to contagious diseases; in one study entire combined litters were repeatedly wiped out when infections spread through, which would probably have not been the case had the kittens been reared in physically well-separated nests.

So it is possible that, like other adult social behaviors of the domestic cat, the communal rearing of young is just one of those things that happens when circumstances conspire to plunk cats down in a novel environment, one very different from the jungles of Africa from whence they came. The ties that cats form in social groups can be deep and lasting, but they are ties that arise from the sum of individuals acting as individuals together.

FOUR

Outta My Face, and Other Useful Expressions

To human beings, communication is a means of expressing ideas. To cats, it is, too. The major difference is that for cats, the principal idea that is usually in need of expression is, "Get out of my face."

The ability of human language to embody an endless number of thoughts, and thoughts about thoughts, is much of what makes humans human. It has also tended to color how we think about communication in other species; it is only natural for us to see in the sounds and gestures of other species a rudimentary system of words and sentences. But this "semantic" view of animal communication—this notion that each *grrr* or whistle or squeak has a precise meaning that can be put into words—has actually been an obstacle to our understanding what is going on when cats and other animals communicate with one another.

From an evolutionary point of view, the only reason an animal would acquire the ability to send a message, or to perceive a message sent by another, is if it confers some survival value for both sender and receiver. Eugene Morton of the Smithsonian Institution, a leading researcher in animal communication, has long argued that the key to understanding how animal communication systems evolved and what purpose they serve in an animal's life is to ask not what a particular sound or gesture *means* but what it *accomplishes.*

The overriding thing that the communication system of most nonhuman animals accomplishes is to avoid bloody fights whenever they can be avoided. As all school yard bullies and international negotiators know, getting what you want through the mere threat of violence is a much better deal than actually having to fight it out. Even the winner of a fight may suffer an injury in the course of battle, so it almost always pays for an animal to issue an unequivocal warning before launching an attack. Likewise, an animal who is able to successfully read and anticipate the hostile intent of another may have time to duck a fight he may not really be after, either because he suspects he cannot win it, or because he in fact has little at stake in the outcome.

Animals that live in large groups often have a very rich and subtle array of sounds and stylized gestures that serve to signal dominance and submission. Social animals that evolved in forested habitats, such as wild pigs, tend to have a particularly rich vocal repertoire; social animals such as horses that live in open habitats, where members of the group are generally within line of sight of one another, tend to have a comparatively richer visual repertoire. Dogs, somewhere in between, are extremely expressive both vocally and in body language (and thus usually have little trouble making even their thickheaded owners know their intent). Dogs in particular have an elaborate roster of submissive behaviors designed to make sure a higher ranking member of the pack knows he is not being challenged. Dogs cringe, whine, drop their heads, lay their ears back, roll over, tuck their tails between their legs; and not only other dogs but most people, too, instinctively know exactly what these gestures convey.

More solitary animals like the cat, however, have a much more limited menu of both sounds and body language; reading the cat's intent can thus be a more difficult proposition. The key lies in understanding the basic evolutionary principles that have selected certain signals for their effectiveness.

Body Language

Cats can and do communicate aggression, fear, and submission through their body language. Yet one reason cats do not communicate submission as emphatically as dogs is that among wildcats, the loser of a dominance encounter generally just buggers off, and fast. This is quite different from the situation with group-dwelling animals that are constantly forced into proximity with one another and so need to be constantly deflecting aggression. Dogs carry on an ongoing dialogue over their relative social positions within the group hierarchy, and in the aftermath of each dominance encounter everyone is still a member of the group and has to keep getting along. If you are trapped in a crowded subway car, it is extremely helpful that there are social conventions such as saying "pardon me" to deal with situations that may arise, such as accidentally poking your umbrella into a 280-pound guy with a shaved head wearing chains and a black leather jacket.

Body language: The arched back of the defensively aggressive cat combines the stiff-legged stalk of the aggressive cat with the crouch of the fearful cat.

In their natural state, however, cats invariably face off only in disputes over territory. A territorial dispute ends, by definition, when one cat leaves the territory. There's no need for further groveling.

As a result, the emphasis in the cat's evolutionary history has been on the development of aggressive signals. In general, aggressive signals are functionally connected to actual aggressive action; they are shows of force, such as the human gesture of shaking one's fist. But there is also a certain arbitrary nature to the process by which certain physical gestures become "ritualized" as symbols, and this is the result of an interesting and powerful evolutionary feedback between senders and receivers. It was not inevitable that a mere look, for example, would end up as a potent symbol of threatening intent, but in the cat world it is. The explanation is this: A cat who is about to attack another cat *has* to do certain things. He has to look at his intended victim, for starters. Cats who thus began to correctly interpret a stare as a sign of an imminent attack enjoyed a competitive edge in the evolutionary sweepstakes—they were more likely to avoid unnecessary fights, and so survive and pass on to their

Some common facial expressions in cats

offspring that tendency to "read" a stare as a threat. (Clueless cats, on the other hand, were more likely to blunder into those unnecessary fights and get injured or killed.) The more that cats became sensitive to *reading* a stare as a threat, the greater the payoff then became in *using* a stare as a deliberate threat. Cats who began to fix their intended victims with a prolonged stare before attacking were thus particularly effective in getting another cat to back down without having to attack; and these stare-as-a-threat cats were thus more likely to survive and pass on this tendency to use a stare as a threat.

Over time a fairly innocuous action came to have a very specific purpose. I recently was given a remarkable book that traces back to Greek antiquity the interesting hand gestures that form such a rich part of the communicative repertoire of Neapolitan Italy. The analogy to the ritualization of animal body language is striking. Some gestures, especially some of the more obscene ones, have a pretty clear connection to their intended meaning. But some are equally clearly the product of pure ritualization over time. My favorite example of the latter—it is featured under the heading, "derision, ridicule"—is a gesture that was derived in the first instance from the practice of making a rude sound by blowing into the hand. The complete technical description of this derived gesture runs as follows: "Palm of the hand placed under the armpit of the opposite arm. The hand is so arranged that, when compressed with violent blows given to it by the arm, because the air trapped there is pushed out by the force of the blows, it produces the same sound as that obtained by the mouth, but even more stridently. More emphasis is given to this gesture by lifting a little the leg corresponding to the arm that presses the hand." In a clear sign of ritualization, the nineteenth century author of this learned work continues, "Even if just the first phase of this gesture is performed, it has the same meaning. This may be done simply by bringing a hand under the opposite armpit, and lifting the corresponding leg a little and adding, further, an ironic expression on the face."

Cats indicate an offensive intent through other ritualized symbols as well. They hold their tails low and close to the body, their legs straight, their hindquarters elevated, and the ears rotated to the side. A relaxed and alert cat, by comparison, stands with his ears forward, his tail hanging loosely and easily behind, and his body level.

For a time, animal behaviorists tended to the view that cats really did not express submission at all. But a recent study by Hilary Feldman of Cambridge University suggests that cats do in certain contexts express submission with a rolling-over behavior that is rather similar to that seen regularly in dogs. Rolling over and exposing the vulnerable abdomen is a classic submissive pose, for it expresses the animal's total passivity and vulnerability. Feldman found that rolling in cats was not random but nearly always directed at other cats. Most of the time rolling was part of male–female courtship behavior. But a not insignificant number of the rolls she observed were between males. Typically a young male would approach an adult male and ostentatiously roll in front of him; if the older cat made any move in response the young cat would freeze with his belly fully exposed. In not a single case did the older male respond aggressively. Feldman suggests that while such submissive behavior does not play a part of normal antagonistic encounters between adult cats, it may have evolved as an important function for the period when young cats are approaching sexual maturity but have not yet dispersed from their mother and littermates, the period around four or five months of age. At that time older male cats begin to view the young males as competitive interlopers, and the young males greatly benefit from having a way to avert hostile action. Significantly, kittens also use the roll with one another to initiate play, much as dogs will "bow down" to one another before play; in both cases it is a way to communicate a nonhostile intent to actions that might otherwise be seen as potentially threatening. House cats appear to make use of this part of their instinctive communicative repertoire to deal with the slightly odd circumstances of living with people; pet cats, even as adults, will frequently roll as a way to initiate play or other friendly interaction with their human owners.

Submission implies the abandonment of any attempt at defense; it is effectively an act of surrender. A cat who is frightened, however, is in a different state of motivation; he is ever ready to defend himself. The basic hostile–defensive posture is one that is purely functional: a crouch, with the head low to the ground, the ears tucked back, the whiskers flattened, and the tail and legs tucked underneath, keeping those vulnerable parts out of harm's way.

This position evolved not so much as a signal as a practical necessity. But over the course of evolution there was no doubt a selective advantage to the would-be aggressor who had the wit to correctly interpret the crouch as a sign that his potential victim might be ready to defend himself.

As a consequence there may have been an advantage, too, for

the fearful cat to advertise his fear. At first blush this might seem contradictory, and that coolly concealing one's fear is a far wiser policy. But communicative signals tend to be selected for honesty over time. The danger of bluffing is that one's bluff may be called. It's sometimes possible to outbully a bully, but sometimes such a policy only escalates a confrontation—a dominant animal might take a cool, aggressive swagger from another as a challenge that he cannot ignore. A fearful cat who shows he may be driven to attack if provoked, on the other hand, is adopting what Richard Nixon once called his "madman theory": If the rest of the world thinks that the guy with his finger on the button is a bit unstable, they will tiptoe around and be careful not to do anything that might upset him.

As the level of fear increases, cats will increasingly advertise the fact that they feel threatened and that they may be driven to lash out. A very frightened cat is in truth driven by both defensive and offensive motivations, and will begin to superimpose on his defensive posture a clearly ritualized aggressive stance. Still crouching, the cat arches his back and lifts his tail straight up or in an arch, and his hair stands on end. This is the classic "Halloween cat" pose. Puffing oneself up like this is a universally recognized signal of threat in the animal world; it is as close to animal Esperanto as one gets. And part of the reason a fearfully aggressive cat uses this particular kind of signal may be that it serves to communicate not only to other cats but to threatening predators as well. Basically such signals have their roots in trying to look big. This originally may have actually fooled a potential competitor or predator; thus many animals raise their hackles or spread their wings or puff up their breasts in such situations.

Over time, though, this has become a ritualized signal of intent, and is accurately read this way across species lines: Looking big is a universally recognized symbol of an animal's intention to act big. (This schema of size symbolism also underlies patterns of vocal communication, discussed later in this chapter.)

Fear is reliably expressed in the cat in another way: the widening of the pupils. This bug-eyed look is of course a familiar one in frightened people. At root this is a reflex reaction to fear that may be just a coincidence of the nervous system's wiring—much as people when frightened may break into a sweat or flush or develop a nervous tic. But the dilation of the pupils of a frightened cat is so marked and consistent that it may in fact be a ritualized signal. Further evidence of this conclusion is that the eyes appear to have a signaling role in offensive aggression and submission, too. An offensively aggressive cat will narrow his eyes in a beady-eyed stare. Blinking or turning away, the opposite of a fixed stare, is used in some submissive contexts by cats.

This, incidentally, may explain the perverse but pretty nearly undeniable fact that cats always seem to plop down with unerring instinct in the lap of the one guest who hates cats. It is likely that what is going on here is that the person who does not like cats tends not to look at the cat, and from the cat's point of view an averted gaze is a sign of nonthreatening intent, if not outright welcome. A visitor who eagerly looks right at the cat is going to be seen as much more potentially threatening, and the cat will instinctively avoid those well-meaning (but misunderstood) admirers.

Dogs have a ritualized expression of baring their fangs to convey a dominance threat, but cats do not. There is probably no particular reason for this, as there is a certain amount of luck of the evolutionary draw that determines which gestures become ritualized. But cats showing extremely defensive, fear aggression will open their mouths and show their teeth in a distinctly menacing fashion. It may be that this facial expression was selected over time for this defensive-rage function because it worked well against other potential predator species. Anyone who has come upon a feral tomcat unexpectedly and been met with a hiss, an arched back, and a bared mouthful of teeth can certainly testify that it works. We'll get back to the hiss in a moment.

The use of the tail as a signal by various species is a complex business, and attempts to develop a sort of general theory of tail movements reveal a good many exceptions to any general rules that one might be tempted to formulate. Many animals raise their tails both in greeting and in aggressive encounters. Cats will raise their tails straight up in greeting, in play, and in clearly friendly encounters; kittens raise their tails when begging for food from their mothers, and nearly all pet cats will do the same at feeding time as their owners are opening a can of cat food or pouring food into a bowl. The reason that cats—unlike horses, dogs, pigs, and goats, among many other animals—do not raise their tails as an offensive threat, however, may be explained by the fact that there is less ritual to the aggressive encounters in cats and more real fighting. A dog can raise his tail as a threat without worrying that it will be immediately bitten off. Encounters between cats in their natural habitat always involve an element of fear and anxiety and unwelcomeness; even a confident aggressor is motivated by a streak of defensiveness that is expressed, motivationally and as a matter of practical prudence, by keeping his tail low.

Tail lashing or wagging has a less direct functional tie to any particular motivational state than does overall tail carriage, and as a result wagging tends to be ritualized to serve a purpose that varies widely among species. In cats, as in horses, but not at all as in dogs, a lashing tail is generally a manifestation of irritation and defensive aggression.

The World According to Smell

For an animal that spends a lot of time trying to avoid members of its own species, smell offers considerable advantages over other forms of communication. Notably, it is the one method of animal communication that can travel over both space and time. Marking an object with one's odor is a way to leave a message that can be read later by an animal that was nowhere around at the time the message was posted. Moreover, it is a date-stamped message; it can provide information about when it was left, a refinement that only written language as a communication medium can equal.

In keeping with their ecology and social habits, cats have a remarkably rich array of odor signals that they can call upon for a variety of functions. Cats appear to use scent marks to time their movements to avoid one another on shared hunting trails, to tell if they are in the territory of another cat, to locate mates in the breeding season, to identify familiar cats, and possibly to navigate in their home range.

Urine spraying, discussed in Chapter 3, is the howitzer of the cat's scent-marking arsenal, but cats have more fine-caliber armaments as well. Studies by Robert Prescott at Cambridge University in the 1970s determined that the tail, forehead, chin, and lips of the cat are well endowed with sebaceous glands that secrete a fatty substance bearing an individually distinctive odor. Cats frequently rub the lips, chin, and tail against inanimate objects to leave an odor mark. The pads of the feet also contain scent glands, and so scratching is another way to leave a scent mark; scratching has the added advantage of reinforcing such a mark with a visual cue as well. Often scratching is focused on a prominent vertical landmark, such as a tree, that they use over and over and the result is a sort of personal territorial marker that is continually reinforced. Cats do scratch to loosen old and worn claws that are ready to drop off as new, razor-sharp ones grow in beneath, but they will also use their teeth to pull off old claws, so it seems likely that scratching per se is motivated in considerable part by its communicative function and not just as a means of sharpening claws.

Studies have confirmed that cats definitely notice and pay attention to scent marks left by others: Wooden pegs that a female cat had rubbed her lips upon were sniffed for a significantly longer time by male cats than were otherwise identical but unmarked pegs. Cats when sniffing such marked objects will sometimes exhibit the "grimacing" expression that animal behaviorists refer to as *Flehmen,* in which the lips are pulled back and air is drawn in. This behavior is peculiar to the cat family and a few other families of mammals, such as horses, cattle, and sheep, and it is designed to detect social odors through the use of a specialized and highly sensitive organ in the roof of the mouth, known as the vomeronasal organ.

Cats frequently rub their foreheads, and to some extent their tails, on other cats and people, and this suggests a rather more complex social function of scent marking. Kittens rub their mothers this way, and subordinate cats in a group will approach and rub against the dominant animal. Females who live in amicable groups regularly rub each other upon greeting. Experiments have shown that even at a few days of age kittens are able to distinguish the individual odors of individual cats. Kittens removed from their nests at this age—before their eyes open—can successfully orient themselves and find their way back to the nest and can correctly pick out their nest from that of a strange mother. If the floor of the pen is washed first, however, the kittens freeze when removed from the nest and appear totally bewildered. So odors are individually distinctive and individually recognizable.

It is often speculated that cats rub on familiar cats and on people in order to place their odor mark on them, just as they mark inanimate landmarks. But an arguably more logical explanation for this kind of rubbing is that in this case the cat is himself trying to pick up the odor of the animal he is rubbing. Especially if the rubber is encountering a socially superior rubbee, there would be an advantage to this: The rubber would acquire a seal of approval, as it were, that would remind the dominant cat on a subsequent encounter that here is a familiar and accepted member of the dominant cat's society. The value to kittens of bearing their mother's scent is probably quite similar; it would unmistakably identify the kitten as hers and reduce the chance of maternal rejection or neglect.

It is also possible that mutual rubbing in effect creates a collective "group scent" that serves a similar purpose of identifying friend or foe for the litter, or social group, as a whole.

This social scent-marking behavior among adult cats is, in any case, another one of those more promiscuously groupish habits of domestic cats that draws upon much more narrowly focused instinctive behaviors that, in their wild ancestor, occur only between kittens and mothers during the few months that kittens live in a family group before reaching sexual maturity and dispersing, or between courting males and females for the very brief time they are in contact during each mating season.

In pet cats, people probably unconsciously reinforce this instinctive behavior even more by responding with rewarding actions such as petting or playing with a cat who initiates contact by rubbing.

Very Vocal Communication

When human mothers speak to their babies, they use a distinctive kind of speech that has been dubbed "motherese." When researchers recorded British, American, German, French, and Italian mothers uttering words of approval, comfort, admonition, or attention seeking to their infants, they found that in spite of many obvious linguistic and cultural differences, the *melodic* patterns of what all the mothers said were virtually identical in any given situation. Whether a mother was saying "That's a GOOD boy" or "braVISSima," she used a high, quickly rising, then just as quickly falling pattern of pitches. When calling the baby's attention to something, the mothers all used a series of short, rapidly repeated sounds each rising in pitch ("see the BALL? see the BALL?"). Telling the baby not to do something was invariably spoken in a series of loud, low-pitched, abrupt sounds. Comforting words were also spoken in a low pitch, but softly and murmuringly.

Professional animal trainers and handlers have been observed to use exactly the same "language" of melodic and rhythmic patterns to control and manage their animals' behavior.

Obviously there is something going on here. People do this instinctively because it works—neither human babies nor animals understand the semantic meanings of the words their mothers or trainers speak to them, yet the message gets across, and they respond appropriately. And the fact that this language works so universally must hold some clues as to how different vocal patterns have evolved in animal communication.

Cats can be very vocal at times, but the range of their sounds is not so great as those of more social animals; and in general they are also much quieter than, say, dogs or birds or guinea pigs. Yet all of their communicative sounds, and their reactions to sounds, follow very closely to the patterns found in "motherese."

Why do these sounds "mean" what they do? Or, rather, why are they effective in eliciting certain behavioral responses? The basic key to unlocking these patterns is the same size-symbolism rule that explains why an animal displaying a threat tries to look big. Big things make low sounds; therefore low sounds (growls) inherently convey a notion of warning or threat. Little things make high-pitched sounds; and so high-pitched sounds (whines) inherently convey a sense of nonthreatening encouragement or acceptance.

This is something deeply ingrained in nearly all mammals and birds, especially in governing the complex social relationships between parents and offspring. The naturally high-pitched sounds of the young serve as a reinforcement of other signals that inhibit aggressive action by adults; that in turn readily leads to the ritualization of high-pitched sounds in many other contexts (courtship, social submission, mothers calling to their young) where the aim is to communicate one's nonthreatening intentions—in effect, animals have exploited the preexisting bias of parents against harming things that make little high-pitched sounds.

There is some intriguing speculation that the use of sounds in this symbolic fashion may go back to the dinosaur age. The lambeosaurs, large land dinosaurs that lived in the late Cretaceous period, had crests that some paleontologists believe may have functioned as acoustic resonators. Acoustic analysis suggests that adults were able to use these crests to produce sounds over a broad range of pitches; other fossil evidence suggests that these dinosaurs had close parent–offspring ties and complex social structures, which would have provided the necessary impetus for developing such social signaling.

The several exceptions to this general rule about pitch and motivation are also remarkably consistent and widespread in birds and mammals. Many animals make nonvocal sounds—that is, sounds that do not come from the vocal cords or which involve resonances that in part carry through the body in certain unusual ways. And these nonvocal sounds generally do not follow the size-symbolism relationship. But these anomalous sounds are very readily accounted for by the fact that they serve purposes that a size-symbolic signal would be ineffective in accomplishing. Chickadees, for example, will growl at each other to convey threats, but a chickadee growl is still rather high-pitched to many larger animals; so when a chickadee wishes to warn off a squirrel that is at the entrance to its nest hole, it makes a hiss—a sound that was probably selected for its mimicry of a snake, which many land mammals have a well-evolved fear of. (Geese also hiss.)

Another category of partly nonvocal sound that is very common in mammals is a low nicker or murmur, which is used both by mothers and offspring to stay in contact while nursing. This is a sound that actually is not a sound at all, but rather a tactile vibration designed to be felt by direct contact. It happens that low-frequency vibrations travel the best through flesh. Sounds that originated in this fashion have, in many species, been ritualized as low, murmuring vocalizations used in many other contexts, frequently as a courtship call by males—this is the case in horses, for example.

All of this may seem like a strange digression into dinosaurs, chickadees, and human baby talk, but these basic principles do a wonderful job of unlocking the key to the cat's vocal repertoire. The cat's most characteristic sound, the meow, comes in a great many variations; and some authors have attempted to catalog them all with a variety of terms (meow, mew, moan, chirr). But really all of these variants lie on a continuum that is governed by—and explained by—the size-symbolism rule. The basic meow is a neutral rising and falling pattern that is rather akin to the attention-getting sound of "motherese." As meows rise in pitch they reflect more of a supplicating or appeasing or friendly motivation.

The other cat sound that follows directly from the size-symbolism rule is the growl or snarl.

And finally, there is a class of sounds that actually combines both "small" and "big" motivations simultaneously. Big, aggressive sounds are not only low-pitched but also as a rule tend to be rough and raspy; this arises because it is a physical fact that when vocal cords are slackened, to produce low pitches, they also tend to vibrate in more complex patterns and produce many overtones—they are "broadband," complex, buzzy sounds. By contrast, little, nonthreatening sounds tend to be not only high-pitched but also pure and tonal; a taut vocal cord resonates in a simpler pattern that has fewer overtones. Because of this association between roughness and low pitch and tonality and high pitch, these tone qualities have become ritualized and thus form a separate scale on the aggression–appeasement spectrum. Animals can to some extent control pitch and tone quality independently, which means they can convey a wide range of finely tuned nuances. In general the rule seems to be that increasing fear is marked by increasing pitch, while increasing aggressiveness is marked by increasing roughness or "bandwidth." Thus just as an aggressively defensive cat will combine the body postures of fear and aggression, so he will combine the vocal characteristics of fear and aggression. The result is what is often termed a "squeak" or "squeal" or "yowl," which in effect is a simultaneous whine and growl. This is a high-pitched but rough sound, and cats will use it when cornered and pressed into a fight or when experiencing severe pain (when it becomes more of a shriek).

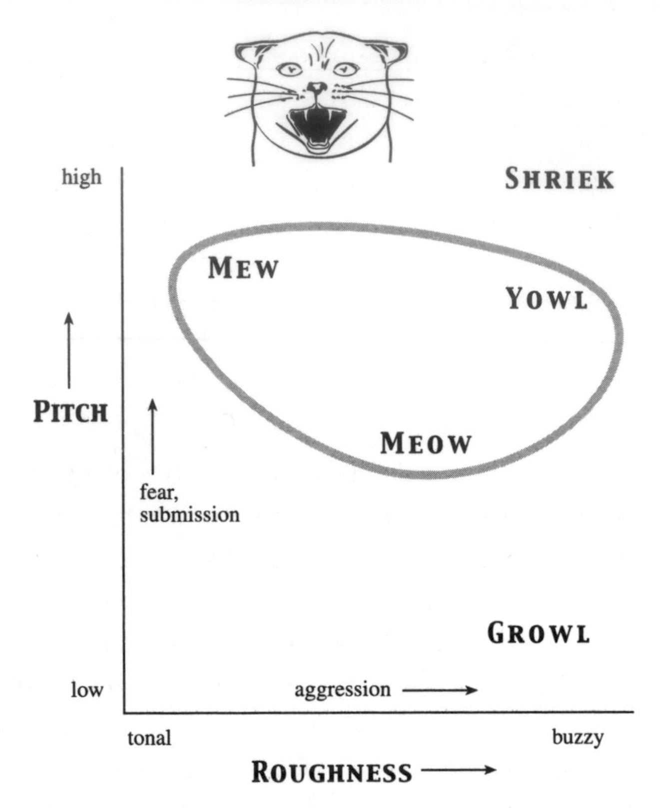

Vocalizations convey information both by their pitch and by their roughness. Pure, high-pitched sounds like the mew are used to indicate fear, appeasement, and submission; a low-pitched buzzy sound like the growl is an unmistakable threat.

Because squeaks combine conflicting motivations in a single sound, they also are sometimes a component of play fighting; in this context they in effect are a threatening growl with a "just kidding" whine superimposed.

Cats make several kinds of nonvocal sounds that also follow the general principles very well. They hiss in defensive situations.

Sometimes the hiss is prefaced by a spit, which is a short, emphatic, and explosive sound that is probably effective because it is simply startling. Cats also produce a low, closed-mouth murmur, sometimes termed a "call," that is used as a social greeting or as a female's signal of sexual receptiveness; this is very much akin to the horse's nicker and is probably a ritualized version of the mother's through-the-body contact call to her kittens.

The Purr

The cat's distinctive purr has long been the subject of a great deal of not always convincing theorizing and speculation. It probably says much more about human ignorance than feline exceptionalism that various recent authors have attributed to the purr such outré meanings as a "mantra" or seen it as a manifestation of the euphoria accompanying near-death experience. Although the cat's purr is exceptional in some ways, its evolution and communicative function are straightforward and fit well with what is known scientifically about other similar, low-rumbling communicative vibrations used by various mammals. Basically the cat's many uses of the purr, like the close-mouthed murmur, represent a ritualization of the nonvocal sound used for maternal, through-the-body communication.

A detailed physiological study of purring by researchers at Dartmouth University's medical school established that purrs are generated by very rapid and very regular nerve impulses sent directly from the central nervous system to the muscles of the diaphragm and to the vocal cords. The muscles are activated alternately in bursts of about twenty to thirty per second. In effect they are extremely well-controlled tremors, and the physiologists who carried out the study noted that no other sorts of normal tremors studied in animals come close to the high frequency and high regularity of the purr.

Significantly, purrs often do not produce any audible sound that carries through the air, though they can always be felt by someone in contact with the purring cat. Kittens begin to purr when a few days old, and do so invariably when suckling; mothers also purr while nursing. That is certainly strong evidence for the purr's basis as a mother–young contact "sound." In its ritualized contexts, purrs are used in courtship, in friendly greetings, and in appeasement of a dominant animal.

Pet cats may be particularly prone to purr when in contact with people because they so readily call upon all the instinctive social behaviors of kittenhood and motherhood in dealing with people. The social tool kit of the cat, including its vocal repertoire, is generally geared for confrontation, not affection. Yet when confronted with man and his comparatively easygoing and noncompetitive ways (easygoing and noncompetitive compared to most cats, that is, and certainly compared to most wildcats in their natural habitat), the pet cat is free to trot out his small, but potent and endearing, repertoire of cooperative and friendly behaviors—behaviors that are all the more potent and endearing for their origin in the bonds between mother and young.

FIVE

The Thinking Cat's Guide to Intelligence

When people who were chosen at random from many different walks of life and educational backgrounds were asked whether dogs or cats are smarter, dogs won hands down.

Cat owners have long had an answer to this popular slander, namely that cats are simply too smart to do all the dumb tricks that dogs so willingly do and which seem to so readily impress us readily impressionable human beings. It is probably giving cats too much credit to ascribe to them the intelligence of having figured out how to beat the system, but there is in fact substantial scientific evidence to support the gist of the cat owners' defense. Human beings tend to be extraordinarily biased in assessing the intelligence of other species. We consistently rate as the most intelligent those animals that, like us, have good eyesight and nimble hands—in other words, animals that can see and react to things that we can see and react to and that can express the results of their mental processes in obvious ways, such as manipulating objects, as we can.

We also have a strong bias toward viewing as intelligent those animals that can learn to do things that we consider useful. But what in truth determines an animal's ability to master a given task often has much less to do with innate intelligence than with behavioral predispositions. Experimental psychologists have learned over many years of trying to measure the relative intelligence of different species that an animal may do extremely poorly at learning certain things, yet when the experiment is redesigned in a way that better suits the behavioral or perceptual peculiarities of the species, the animals ace it. Species differ in how well they can see or hear the visual or auditory cues they are being taught to respond to; they differ in the rewards they are willing to work for; they differ in the things that they are likely to be wary or frightened of, and so interfere with learning. Just as IQ tests have rightly been condemned for being culturally biased, so many of the formal and informal measures of animal intelligence are species biased. It makes as much sense to consider a cat dumber than a dog because cats cannot be taught to fetch as it would make sense to consider an American dumber than a Frenchman because he cannot pass an IQ test written in French.

When it comes to intelligence testing, cats are particularly difficult research subjects. It is hard to get cats to show what they know, especially in a laboratory setting. Social rewards and punishments that are highly effective in teaching dogs and horses are almost meaningless to cats. Cats may *like* to be petted, but to a social animal like the dog, petting carries the powerful significance of social acceptance by a social superior, something that is a matter of indifference to the cat. Punishing a dog or a horse likewise carries the sting of social disapproval, which is many times greater than the physical pain involved; indeed, that is why speaking harshly to a social animal is usually as effective as—or even more effective than—hitting him. A cat, however, normally responds to punishment by fighting back or running away. There is simply little in the cat's social and behavioral makeup that makes a cat want to stay and take it and alter his behavior in response to social pressure, one way or the other.

Even food rewards often seem insufficient to motivate a cat to learn a specific, focused task. As solitary hunters, cats seem to be able to go for long stretches without eating; when placed in a strange situation, they accordingly seem to be able to suppress their hunger more readily than dogs or monkeys or rodents. In one experiment, in which cats were tested for their ability to locate an object hidden behind a screen, the researchers noted that the cats' searches were "slow" and "lackadaisical"—even when they had been deprived of food for twenty-three hours before the test, and when the reward for finding the object was the cats' favorite treats (cooked chicken for one cat, commercial cat treats for another). It is hard to imagine a dog being "lackadaisical" under similar circumstances.

Yet cats clearly do learn in their natural environment, altering their behavior and strategies in complex and often novel ways to changing circumstances. And cleverly designed experiments that take into account the cats' physical and behavioral biases show them unquestionably to be every bit as intelligent as other domestic species.

Input and Output

A large part of what determines how a particular species' innate intelligence will reveal itself is simply what kind of sense organs and motor abilities it has. If one looks at how the brains of various mammals are "wired," a major organizing principle is the relative role that their various organs and limbs play in their lives. The neocortex—the sheet of "gray matter" that surrounds the core of the brain and which plays a crucial part in learning—is highly specialized according to species. New World monkeys that grasp with their tails have a particularly large chunk of the neocortex devoted to controlling the muscles of the end of the tail. Humans have a large visual area in the neocortex and also a large part devoted to controlling fine muscle movements of the hands. Cats make particularly heavy use of their sense of hearing to locate and hunt prey in the hours of dim light when they are active at dusk and dawn, and they have a correspondingly large auditory portion of the neocortex.

Perceptual abilities also of course determine much about how an animal understands and orders his world; they are the raw inputs that its intelligence can act upon. Humans have a huge bias toward the visual, and we have an extremely difficult time even imagining what the world would feel like to a being with a different set of perceptual abilities. We tend to think of our pets as seeing things more or less as we do, but that is not the case at all.

To begin with, cats, like nearly all mammals other than the primates, are color-blind. The issue of color vision in cats has been remarkably confused and controversial over the years, with some studies seeming to show that they have full color vision akin to that in human beings. But definitive studies, which have measured electrical nerve impulses in cats as they look at various colors and tested which colors cats can be taught to distinguish between in learning experiments, have shown otherwise. The cones, or color-sensitive cells, in the human retina come in three varieties, each of which is most sensitive to light of a particular wavelength—roughly those corresponding to the colors red, green, and blue. By monitoring the relative strength of the nerve signals emanating from each of these three kinds of cones, the human mind is capable of calculating the wavelength of all the in-between colors in the spectrum as well, the yellows and oranges and violets. Human beings can distinguish about a hundred different distinct hues in laboratory experiments.

Cats, however, pretty clearly have only two kinds of cones, and these are most sensitive to green and blue light only. That means that to a cat, the colors red, orange, yellow, and green are all really just one color, the various shades of blue and violet another. Cats can see two colors, in other words. They know that a red ball is not gray or white or black; but they can't tell the difference between a red ball and a green ball.

Part of the evolutionary reason for this is that there is a trade-off between being able to see well in color and being able to see well in dim light. The rod cells of the eye—which are sensitive to overall light levels but not to any particular color, and which thus can give a reading only along a black–gray–white scale—are much more effective in low-light conditions. Color vision is particularly important to animals that eat fruits, such as some primates and birds, and to animals that need to be able to spot camouflaged predators. Many birds, reptiles, and fish have full, three-color vision, and some fish and birds even have four-color vision: They can see into the infrared spectrum. But mammals probably lost three-color vision early in their evolutionary history, as the only niche available to the first mammals, in the age of a lot of big and predatory dinosaurs, was the nocturnal one. There is a fixed amount of space on the retina available for nerve cells, so an animal in effect has to choose between rods and cones, and being active at night would certainly have placed a premium on rods. Giving up some color vision was apparently a reasonable price to pay. Cats appear to have made the low-light/color trade-off a bit more than dogs; their sensitivity to color does not seem as good as that in many other species with two-color vision.

Two-color vision is a good compromise because it still allows an animal that is active during at least part of the day to avoid being duped by rudimentary camouflage of either predators or prey. The three-color vision of man and some of his fellow primates was probably a much later "reinvention" of a trait that was lost long ago in the course of mammalian evolution.

Cats, like other mammals that are most active at night or in the hours of dim light, show other adaptations in their visual system. Behind the retina is a reflective layer of tissue called the tapetum lucidum that in effect gives the retina a second chance to detect an incoming photon by bouncing it back through the retinal cells. The tapetum lucidum also has the incidental property of causing a cat's eyes to glow yellowish when caught in the headlights of a car or the flash of a camera. (The "red eye" of human subjects captured on flash pictures is the result of the much less intense reflection of light off the blood vessels that twine through the back of the retina.)

Cats differ from man also in visual acuity, that is, how sharply they are able to see things. Sharpness of vision is a function of many factors, including the eye's ability to focus, the size of the eye, the density of light-sensitive nerve cells on the retina, and how many retinal cells feed into each of the larger "trunk" cells that relay the visual signals on. There are trade-offs here as well; feeding several retinal cells into one trunk cell increases sensitivity to very low levels of light but at the same time makes for a coarser image. This is exactly analogous to the trade-off in photographic film between sharpness and speed: Big globs of light-sensitive chemicals on the film respond to lower levels of light but make for a grainier image.

Human beings have extremely good visual acuity; a person with normal vision can pick out an alternating pattern of black-and-white stripes when each stripe fills as little as one-sixtieth of a degree of arc in his visual field. (Beyond that point, the stripes blur together and the pattern looks like an even gray.) Doing this test in animals is a bit more of a challenge, but reliable results have been obtained by measuring brain-wave patterns, continually narrowing the stripes until the signal from the animal's visual cortex undergoes a characteristic change; or by training an animal to consistently nose a card bearing a pattern of stripes in preference to a gray card, and then narrowing the stripes until the animal's ability to pick the right card suddenly drops to the fifty-fifty chance level. Such experiments have shown that cats' visual acuity is about four to ten times worse than that of humans; this corresponds to something around 20/80 vision, meaning that what a normal sighted person could see at 80 feet, a cat would have to be 20 feet from to see equally well. Dogs are slightly better than cats, but in the same ballpark. (Eagles, by contrast, can see details about four or five times better than people can.)

Because cats' eyes are mounted in a very forward-facing position on the skull, their ability to judge size and depth—which requires the field of view of both eyes to overlap—is extremely good. In effect, the brain uses the small discrepancies between what each eye sees when looking at the same object to ascertain the true distance, and thus true size, of that object. By contrast, grazing animals, which are also generally prey animals, have side-mounted eyes that give them an extremely wide field of view; horses, cows, and sheep can see almost completely behind themselves. This allows them to detect anything sneaking up on them in any direction. But it also limits the overlap of the left and right eyes to a very narrow band, if any at all. Carnivores trade off a wide overall field of view for a wide binocular field of view. The stalk-and-pounce method of hunting of the cat requires extremely fine judgments of distance, which may explain why cats tend to have a considerably greater field of binocular vision than do dogs, about 90 to 130 degrees depending on the individual, as compared to 60 degrees for the typical dog. (Humans come in at 120.)

Cats, like many mammals, use their extremely sensitive whiskers to supplement their sense of vision. Cats can feel their way over and around obstacles with great precision even when

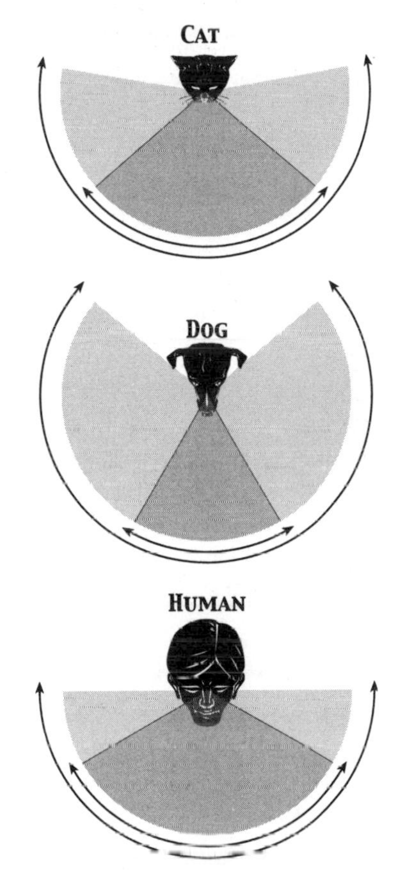

Cats, like people, have very forward-facing eyes. That reduces their total field of view (light shading) but maximizes the region in which both eyes work together (dark shading).

blindfolded by using feedback supplied by their whiskers. The nerves connected to the whiskers, hairs, and other touch-sensitive parts of the body feed into nerve cells of the cortex to form a sort of brain "map" that replicates the geometry of the cat's body—though with significant distortions: In cats, as in most bewhiskered mammals, the nerves devoted to the whiskers occupy a disproportionately large territory of the brain map. (In humans, the nerves devoted to the hands and fingers dominate.)

Cats do not have a particularly great sense of smell; dogs leave them in the dust in that department. But they do have superb hearing. They can hear sounds up to about 65,000 hertz (Hz), or cycles per second, far above the absolute limits of human hearing, which are about 20,000 Hz. The only animals that do significantly better are bats and some insects such as moths, which can detect sounds up to around 100,000 Hz. Cats themselves do not emit noises in this

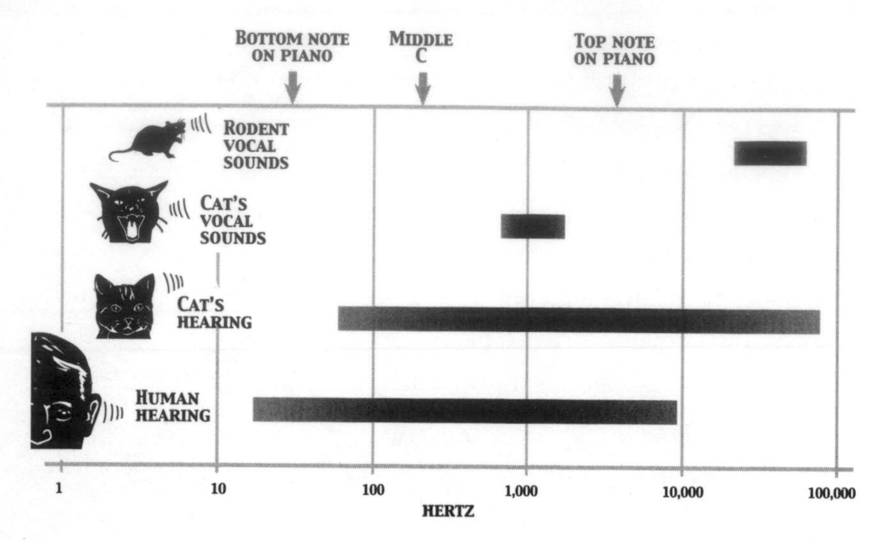

Cats can hear much higher-pitched sounds than people can, an adaptation to hunting rodents.

ultrasonic range, but mice and rats do; and so the cat's ability to hear in this range is almost certainly an adaptation driven by its usefulness for finding and locating prey.

Cats and dogs have about an equal ability to determine the direction a sound is coming from—they can pinpoint it to within about 8 degrees of arc, far better than many animals. Since ability to locate the direction of a sound is in part a function of the distance between the ears, the fact that the considerably smaller cat can equal the performance of the dog, and outperform the much larger horse and cow, means that it has much more intricate brain wiring devoted to the task.

Cats also have some intricate brain wiring for control of their paws. Photographic and X-ray studies of cats' paws in action have shown that they use a variety of dexterous techniques for seizing and manipulating objects. Depending on its size and shape, the object may be pierced with just the claws, held between a claw and pad of the paws, or sometimes held between the paw pads without the use of the claws at all. Many cats also can move the digits of their paws separately, at least to some extent, which implies a degree of fine neuronal control that is probably lacking in dogs and many other quadrupeds. Cats as they are reaching out to grab something "preshape" their grip much as people do, and then rapidly close the paw just as they are about to make contact, indicating that far more than just a mechanical reflex action is involved.

Hard Wiring and Instinctive Intelligence

All animals probably share certain kinds of general intelligence, such as an ability to learn through experience. But they also have their unique, species-specific kinds of intelligence. To some extent, as I mentioned, this species-specific intelligence is just a reflection of the relative importance of the various sense organs. But there are also deeper differences that make it very hard to say whether one animal is smarter than another. In many species there seem to be specialized modules of the brain optimized to perform certain tasks of biological importance for that species' unique way of life. Some species of birds that cache seeds for the winter have phenomenal spatial memory (and a correspondingly large hippocampus, the part of the brain devoted to spatial sense). Other birds are remarkably good at categorizing and remembering the different wing markings on various species of moths, some of which are edible and some of which are not; most human subjects would probably have a hard time equaling the moth-categorizing skills of some jays. Human beings have a highly developed "language module" in the brain that permits infants to perform the astonishing feat of acquiring a language, with its myriad complex rules of grammar, just by listening to it spoken.

I include this last example in particular because it is a good illustration of a fact often overlooked in considering animal intelligence: Just because something the brain does is instinctive, universal, and apparently effortless does not mean that it is any less a part of intelligence than the things that require a lot of groaning, furious scribbling on pieces of paper, and cups of coffee. We don't normally think of it as a sign of particular brilliance that a baby learns to speak English without any formal instruction; all babies who grow up around people speaking English do that. Yet language ability is clearly a major component of the overall intelligence of our species and part of what sets humans apart from other animals.

A good way to attain the proper perspective on this is to ask how hard it would be to build a computer that duplicates a particular mental feat, regardless of whether that feat is instinctive or learned. It is interesting that many of the intellectual accomplishments that set one human apart from another—the things we admire and value because they require work or special talent—are relatively easily programmed into a computer. People have built computers that can beat the world's human chess champion and computers that can solve mathematical equations faster than any human being on Earth. People cannot, however, as yet build a computer that learns a human language spontaneously, or even a robot that can run across uneven ground without falling over flat on its face. In terms of processing and analyzing information, these tasks in fact reflect a huge amount of computing power—and intelligence—however "instinctive" and universal they may be among human beings.

So, likewise, there are many things that cats do instinctively that reflect an extraordinary degree of brainpower at work, even if we don't usually think of them as being signs of intelligence. Much of what makes various breeds of dog seem "smart" to us, for example, are the particular innate mental qualities of breeds that have been emphasized through artificial selection over time. The herding instincts of a Border collie is unquestionably intelligent behavior, involving many rapid-fire calculations; so too is the scenting, trailing, and cooperative behavior exhibited by scent hounds such as foxhounds and beagles that work in groups; so too the retrieving behavior of bird dogs. But that does not make these dogs "smarter" overall than the mutt who can find his way down a complex series of trails, optimize his foraging of garbage cans according to which is likely to have the most goodies on a given day, recognize and know the social status of dozens of other dogs by sight and smell, or follow the trail of a rabbit that passed half an hour before.

There has been almost no selective breeding of cats for behavior, much less for such specialized forms of instinctive intelligence that we see in working dogs; and so there is little that leaps out to the casual human observer as obviously "smart" about many of the instinctive things cats do.

Yet certainly one of the most remarkable manifestations of instinctive intelligence in cats is their hunting ability. The ability to follow the motion of fast-moving prey and to coordinate the movement of the paws and mouth to seize it is an amazing cognitive feat (again, imagine trying to build a robot that duplicates this). Although there is a large component of learning involved in shaping the hunting ability of cats, the basic skill is innate in the specialized wiring of the cat's brain. Cats who have *never* hunted will exhibit the sequence of springing upon and gripping prey if the correct spot in their brain is stimulated with an inserted electrode. (The fact that there is such a specialized, hard-wired pathway for hunting in the brain of the cat helps to explain why cats will often hunt and kill prey regardless of how hungry they are. In most cats, the stimulus of seeing a prey animal triggers a predatory response that verges on the automatic and uncontrollable. Cats are even more natural-born killers than many other predators because they have evolved to be "opportunistic" hunters whose strategy is to make frequent small meals throughout the day from whatever prey is available. It may just be a coincidence, but cats who are given free access to cat food typically will eat small meals that are each just about the equivalent of one mouse in size.)

Part of the innate hunting intelligence of cats is an ability to instinctively differentiate between the motion of a living thing like a mouse and the motion of an inanimate object like a leaf blowing in a breeze. In a fascinating experiment, cats were shown two screens, one containing fourteen computer-generated dots that represented the outline of a walking or running cat, and the other an equal number of randomly moving dots. The cats were able to consistently distinguish the two patterns, but when the "biological motion" screen was turned upside down, they could no longer distinguish it from the random pattern. This ability to recognize plausible biological motion is an extremely complex computational feat that would be a challenge to program into a computer.

Another behavior that is a good example of the specialized, instinctive intelligence of cats is their justly famous air-righting reaction, the ability to turn over while falling so as to land on all four feet. In a careful series of measurements that involved dropping upside-down kittens from a height of 40 centimeters (16 inches), researchers found that kittens at about four weeks of age essentially had no ability at all to turn over in midair, but gradually over the next two weeks they got better and better and by six weeks were hitting the ground with all four feet consistently. (The researchers, no doubt fearing a visit from the Cat Liberation Front, were careful to note that they used a padded surface for the tests.)

The sort of hunting that cats do, however, may have endowed them with somewhat less of one kind of innate brainpower that dogs, and indeed many herbivores—and possibly even some insects—possess. This has to do with the general area of spatial intelligence. Animals whose approach to food gathering requires them to revisit and search a series of fixed sites generally seem to have built-in brain circuitry that not only allows them to find their way back to specific spots but also to optimize their route in order to concentrate on the sites that pay off the most. Nectar-feeding birds, for example, will not return to a flower they have recently visited but will wait several days for it to refill

The air-righting reflex is a complex, but completely instinctive, motor pattern that develops in kittens between the ages of four and six weeks.

with nectar; vervet monkeys will choose a route for foraging trips through their territory that takes them to the richest sites first; dogs will decide which of several scavenging sites to visit based on a rather fine calculation that draws on past experience to determine which offers the highest probability of paying off.

But as opportunistic hunters, cats have the luxury of just taking whatever they find; they have little need for elaborate search strategies and probably do not even set out on a hunting expedition with any deliberate decision of whether to hunt mice or birds or rabbits. That is not always the most efficient strategy, but it seems that that doesn't really matter. Although the amount of time individual cats spend hunting varies widely, studies suggest that most cats, even feral cats who have to fend for themselves entirely, spend only a few hours a day at it in all, and each hunting excursion is typically less than 30 minutes. Laboratory experiments in which cats have to learn and remember certain kinds of spatial relationships—which I will come back to later in this chapter—do seem to show that this is not something that comes all that naturally to cats. The short explanation for this deficit is that they can probably get along just fine without it and have other more useful things to put their brains to.

Although cats do not perform these sorts of spatial calculations as well as some species, their decisions about how much time to invest in hunting do appear to reflect a calculation that integrates a considerable amount of time information. Interestingly, cats are able to discriminate time intervals with an impressive degree of precision. In one experiment, cats were placed in cages for either 5 seconds or 20 seconds and when released were rewarded with a food treat that would always be hidden in the left-hand feeder if they had been in the cage for 20 seconds and in the right-hand feeder if they had been in for 5 seconds. If the cat went to the wrong feeder, that was scored as an error. After 400 to 1,600 repetitions of the drill, depending on the individual cat, all fourteen cats tested were able to pick the correct feeder consistently at least 80 percent of the time. The experimenters then started shortening the 20-second trials to see if the cats could still tell the difference; half of the cats were able to consistently discriminate a 5-second interval in the cage from an 8-second interval.

In another experiment cats had to press a bar a number of times to gain access to food; they could then eat all they wanted at that sitting. As the number of bar presses required for the food tray to open was increased (from 40 to 2,560), the cats responded by eating fewer meals each day, but eating more at each sitting. But what was most interesting was that when the number of bar presses was varied from one meal to the next, the cats tended to adopt a strategy that very closely reflected the average "price" per meal. In other words, they did not adjust their schedule of eating or the amount they ate at a given meal according to how many times they had to press the bar that particular time, but according to the average number of times they had to press the bar for each meal over the course of a whole day or several days.

This ability to calculate meal frequency versus average time expenditure fits well with the sort of real-world problems a cat faces in pursuing an opportunistic hunting strategy. When it comes to finding food, cats are good at time because time matters; they are not so good at space because space doesn't matter.

Learning and Limits on Learning

When cats can be coaxed to show what they can learn—which, as I have said, is not always easy—they do quite well on the classic kinds of learning experiments that psychologists have traditionally used to measure the intelligence of animals.

In one set of experiments, cats are presented with a pair of wooden figures that differ in several marked ways from each other, such as shape, color, and surface area—they might be a black square and white triangle, for example. The cat gets to indicate his choice of one of the two objects by nosing it. Every time he picks, say, the black square, he gets a piece of meat as a reward. Every time he chooses the white triangle he gets nothing. He is given the same problem over and over (the experimenters randomly change the left and right position of the objects each time so the cat doesn't learn to pick the object on one side), and after a certain amount of patient repetition, just about all cats will, unsurprisingly, get the hang of it.

The interesting thing, though, is that cats can learn not only the correct answer on a single problem of this sort but also they can actually learn to generalize from the experience. Once they have a couple of these "object discrimination" problems under their belts, they catch on faster and faster to each new problem of the same kind. At first, they need dozens of repetitive trials with each new pair of objects before they get to the point that they are picking the right answer 80 percent of the time. By the time they've mastered sixty different problems of this kind, they will often hit the 80 percent mark in the first ten tries at it. In other words, they have learned the rules of the game—they have learned that whenever two new objects are presented, the name of the game is to figure out which one produces the reward.

Interestingly, cats are not quite as quick to extrapolate from a *wrong* answer as are rhesus monkeys and chimpanzees who have been given similar tests. If the cat doing these experiments is lucky enough to pick the right answer the first try, and so earn a reward, he masters the new problem far faster than if he picks the wrong, unrewarded answer the first try. The reason for this again may have to do more with ecology than pure intelligence. A cat who is searching for food in the wild is rarely faced with a real-world situation in which if a mouse is not found at place A he will be at place B. But foraging animals like primates who repeatedly revisit fixed food sources such as fruit and nut trees are always making choices about which one to try. So being able to extrapolate from the experience of *not* getting a reward may simply be a more "natural" problem for a foraging primate than it is for an opportunistic carnivore.

It is also a fact that cats are by temperament simply quite averse to putting up with frustrating situations. They tend to become lackadaisical or indifferent when faced with situations that yield no clear path to a reward. Cats will readily "learn to learn" when rewarded for their efforts in a clear and effective manner, but by the same token they can actually learn *not* to learn—or not to bother learning—when faced with problems that do not seem to offer straightforward rewards. Cats have a generally tougher time with a problem in which they have to learn to pick an object on a given side, left or right, depending on which of two possible pairs of identical objects (two white squares versus two black triangles, for example) is presented. This is an even more ecologically unnatural problem for a cat, and so it's not surprising that they have a harder time grasping what's expected of them. Most cats do eventually catch on to such tough problems to the point that they perform better than chance, but not much better; and they do not show anything like the rapid improvement that they achieve when tackling a series of the simpler object-discrimination problems.

But the really telling point is that cats who are given a mix of simple and tough problems catch on faster to the tough problems than do cats who are given a straight course of nothing but the tough problems. One cat who had been in the "tough-only" class, doing nothing but these hard problems, could *never* subsequently be taught to master even a simple black–white discrimination task, even after 600 trials. Basically the cats who are in the tough-only group simply give up, and content themselves with getting a random handout every now and then when they choose the right answer by chance.

While this to-hell-with-it attitude may not reflect on their true intelligence—cats might be the original underachievers—it does have practical consequences for anyone who wants to teach things to cats. All animals are predisposed ("prepared," in psychology lingo) to learn certain types of associations and are likewise predisposed not to learn others ("contraprepared"). They are predisposed to interact with their environment in certain ways, to pick up on certain kinds of cues, and to manipulate objects or their bodies in certain ways. And they are predisposed to work for certain kinds of rewards. Dogs readily work for social rewards such as attention and petting because that is part of their basic social structure. They readily learn to do things like lie down and roll over because those are naturally subordinate behaviors among dogs that they show to their social superiors. They are good at learning to follow scents, to carry things in their mouths, and to herd livestock because those are likewise basic parts of their brain wiring.

Cats on the other hand react badly to social punishment, which provokes a defensive confrontation or flight; they react badly to problems in which the reward does not come right away, which quickly engenders indifference; they react badly to spatial learning tasks. The point is that learning is a matter of smarts but it's also a matter of temperament and personality. In some of the early classical psychological experiments done on cats, they readily learned to escape from "puzzle boxes" by manipulating strings or levers in certain sequences, and they learned to learn, getting better and faster at mastering subsequent problems of a like nature; but they could never learn the secret of getting out of the box when the experimenter opened the door to the box only when the cat scratched or licked himself. Associating an instinctive manipulative action like pawing an object with some external real-world consequence is something the cat's brain is set up to learn; it's natural. Associating an instinctive grooming action like licking or scratching with some external real-world consequence is bizarre and unnatural, and cats cannot learn it.

Psychologists had once believed that the ability to learn is something that appears generally later in life among animals like cats that are born in a helpless and dependent state. But exhaustive developmental studies in kittens have found that, to the contrary, the innate mental ability to learn is present from the start. Kittens even a few days old can be trained by reward to preferentially suckle from one of two artificial nipples distinguished by texture, location, or odor. Kittens learn by the end of their first week of life to distinguish by odor the home region of their cage from other parts of the cage. The conclusion is that what determines learning ability even in kittens is not so much innate brainpower as behavioral development. It is the ability to take in information and to do something useful with that information in the real world, it is the predisposition to find certain things in the world important and certain associations important, and it is the temperament and motivation to explore and master certain things and not to withdraw from them that are the most important factors in determining whether that underlying and ever-present ability to learn is brought to bear.

One lesson of practical consequence from this research on learning in cats is that early experience interacts with natural instinct and behavioral development to shape the future behavior and lifelong learning ability of the cat. Cats who learn not only learn but also learn to learn. Cats with certain personality types—which in turn are significantly influenced by what they learn socially in early life—are also more motivated to learn and more temperamentally able to learn. This is a topic for the next chapter, where I discuss individual personality and how owners can influence it, especially in kittenhood.

Can Cats Learn Tricks?

Another practical consequence of this research is that it is quite possible to teach cats tricks—far more possible, in fact, than most cat owners generally believe. The laboratory experiments on cat learning ability show that they clearly possess intelligent and keen minds. But the trick to teaching cats tricks is that they have to be approached much more in the manner that professional animal trainers use with wild animals. Most of the things people might want a cat to learn are not things any cat is likely to exhibit spontaneously and completely by chance right off the bat, the way most dogs will do the standard dog tricks spontaneously (lying down, shaking hands, speaking, sitting, fetching thrown toys, coming when called) and can then be easily rewarded for doing so in association with a spoken command. You throw a stick for a dog; the dog runs and grabs it; and if he happens to come back to his owner with it and is praised for that, that's almost all it takes for him to learn a whole routine associated with the command "fetch."

In the case of cats, the most effective procedure is one that animal trainers call "shaping." The idea is to break down the teaching process into very small incremental steps, each one of which is easily mastered. At every stage, the key is to set up the training session so the cat has a very high probability of doing the right thing, since punishment for an incorrect response or even a lack of reward for an incorrect response causes the cat to become frightened, mad, or bored with the whole exercise. To start with, one looks for something that a cat is likely to do spontaneously that at least vaguely resembles the final, desired behavior. The veterinarian Victoria Voith described how a colleague of hers taught her cat Sebastian to jump over a fly swatter using this approach. First the fly swatter was just put on the floor, and when Sebastian happened to walk over it, he was given a piece of cat food. After Sebastian did this regularly whenever he was confronted with a fly swatter, Voith's friend started holding the fly swatter up a few inches and again rewarding Sebastian for stepping over it. Eventually Sebastian had to start leaping over the fly swatter.

Likewise, Sebastian was taught to shake hands and roll over on command by such a process of successive approximation and reward. Again, the key was to set things up so that at every stage Sebastian would almost always get the "right answer," and thus a reward—by starting with something easy and virtually automatic, and then shifting the right answer very gradually and smoothly toward the ultimate goal. Voith's friend began by rewarding Sebastian for lying down when gently pressed on the back between the shoulder blades. She then started making the push gentler and gentler until it was just a hand signal. She then started pairing a voice command with the hand signal, again rewarding Sebastian every time he would lie down, until he would do so to the voice command alone. Finally she began manually rolling him over after he obeyed the lie down command, now rewarding him only when he was turned on his back. As before, she began using a gentler and gentler push each time to get him to roll over until finally he would execute the whole routine in response to a single command.

At each stage when the ante is upped, the cat is faced with a sort of either-or choice, with one alternative (the new added bit) rewarded and the other (the old routine) no longer rewarded. But the shift is so gradual that the cat doesn't really have the chance to get frustrated or make a lot of wrong choices.

Once a cat has learned the complete trick, the most effective strategy is to begin to reward the performance only intermittently, not every time. This is a basic psychological principle: Intermittently rewarded behaviors will be retained more consistently when the rewards are withdrawn than behaviors that have been rewarded every time. In a sense this is a sort of shaping, too; another way of looking at it is that it is exploiting the cat's ability to learn a more global, general rule. If the cat learns that the rule of the game is do X and you always get a piece of Tender Vittles, then the first time he does X and does not get a piece of Tender Vittles he may believe the rules of the game have abruptly changed. On the other hand, if over time and a lot of experience he learns that the rule is do X and you *may* get a piece of Tender Vittles, he is far more likely to keep doing it even without an inevitable reward—so long as X has been fully mastered first.

Other tricks that cats have mastered through the shaping of instinctive behaviors include fetching an object and dropping it into a pan, and using a toilet. The veterinarian and animal behavior expert Benjamin Hart has described the process of toilet training cats in several scholarly books and articles. One popular method is to attach a piece of plastic-covered cardboard underneath the regular toilet seat, holding it in place with wires, so that the seat becomes a well in which litter can be placed. Over time the litter is gradually removed and then holes are made in the cardboard, and finally all of the litter and the cardboard is removed. Usually the cat will learn early on to stand with his feet on the rim of the toilet seat to steady himself, probably because the plastic-covered cardboard has an unsteady feel to it. By the time the cardboard is removed the cat has thus learned to jump up on the seat and hold on while doing his business. Hart cautions that the cat "is likely to slip off the toilet seat at least once and fall into the toilet bowl," which will then necessitate backtracking in the training until the cat regains his confidence.

Though cats like dogs can learn tricks, the adage about old dogs and new tricks applies to cats, too. A condition very similar to that of Alzheimer's disease has been identified in geriatric cats. Clinical signs include disorientation, compulsive behavior, disruptions to the sleep cycle, and incontinence. On a molecular level this "cognitive dysfunction" syndrome in cats exhibits a pattern very similar to that of Alzheimer's, too, with plaques of a chemical called β amyloid appearing in the brain. This chemical interferes with the normal action of neurotransmitter chemicals in the brain that relay nerve signals; even worse, it is directly toxic to nerve cells themselves, causing them to die.

Even absent the drastic deterioration that occurs in this Alzheimer's-like disease, old cats learn more slowly as an inevitable consequence of the natural aging process. Studies have found that cats older than ten years are often incapable of acquiring even "classical" or Pavlovian learned associations that young cats learn with no difficulty. These are associations in which a novel stimulus like a sound is paired with a stimulus that triggers an automatic reflex reaction; after repeated exposure to this pairing the sound alone should trigger the reflex. Tests showed old cats were just as awake and alert during training sessions as young cats, and their perceptual nerves were supplying the same inputs to their brains; it's just that the brains weren't running at full steam anymore.

Higher Thoughts

The things that cats do and do not learn, and the ways they do and do not learn them, tell us something about how the cat's mind perceives and orders his world. Learning experiments are the traditional stuff of comparative psychology largely because of the influence of the behaviorists like B. F. Skinner, who thought that all behavior in animals could be explained as the product of simple, learned associations. Behavior was all a matter of nerve cells forging connections between a stimulus and a response, Skinner argued, and even the most complex behaviors exhibited by human beings were nothing but stimulus–response associations. The strict behaviorists were inclined to dismiss thinking and consciousness altogether as a mere illusion.

In more recent decades there has been a growing recognition that a lot of things animals (and people!) do cannot be explained without invoking some higher thought processes. Most comparative psychologists today take a much more cognitive approach to behavior than Skinner and his followers did; they acknowledge that, in carrying out even many of their ordinary daily tasks, animals cannot simply depend upon a rote-learned reflex but must be consulting stored mental images or other sorts of catalogues in their brains to make decisions.

A few experiments in cats have probed the question of what sorts of "mental representations" cats are capable of storing away;

and these I think are particularly interesting in giving us a glimpse beyond just what a cat's mind can *do,* and into the realm of what a cat's mind is *like.* We can never really know what it is like to be a cat without being one ourselves—and actually experiencing the flow of visual and sound signals and the special sense of what matters and what doesn't in the cat's perceptions and calculations. But probing some of the higher-order calculations that go on in the cat's mind is at least a peek at the cat's sense of the universe.

French comparative psychologists, who have long been influenced by the theories of the developmental psychologist Jean Piaget, have never taken quite so didactic a view of learning as the American behaviorists; and one longstanding interest among Piaget's followers is the ability of various species to develop what Piaget termed "object permanence." Piaget noted that human infants go through various stages of understanding about things in the world. At first, human infants show little interest at all if an object like a toy ball is hidden and make little effort to search for it; as soon as the ball is out of sight it no longer seems to exist in the infant's mind at all. At a later age they will search for something that partially or completely disappears but may not understand where to look. Even if they see someone hide the object behind a screen, they will not know to look behind the screen but may, for example, instead look in a place where they previously found it. Still later they will know where to look when an object disappears, and finally they will be able to follow even a series of "invisible displacements": If someone takes a ball, hides it in an opaque cup, carries the cup behind a screen and secretly removes the ball, and then brings the cup back out and shows that it is empty, the infant will know that the ball is behind the screen.

This final stage of object permanence, which Piaget termed Stage 6, emerges in human infants at about eighteen months of age.

Experiments in cats had at first seemed to suggest they never achieve Stage 6 object permanence, but recently more careful studies that adjusted for the cat's particular sensory and ecological biases have shown that to be incorrect. In these tests the cats were tested in their familiar surroundings (their homes, not a laboratory). The screens that were to be used for hiding the object in the actual test were left around for a week so the cats could get used to them, and that also gave them a chance to learn that screens do not as a rule have toys hiding behind them. The cats were first taught that whenever they touched their noses to a particular toy they would get a food treat. For the actual test, two screens were positioned in front of the cat, who was lightly held by his owner while the experimenter put the toy in a cup, secretly removed it behind one of the screens, and then placed the empty cup in front of the cat, who was then released. In nearly every trial, the cat when released went directly behind the screen where the toy had been hidden. The screens were moved from trial to trial and replaced with new screens of a different appearance, and the cats still got the right answer, proving that they had not just learned a "local rule" but had generalized the solution.

To solve these sorts of problems, cats cannot merely learn a rote answer but must consult a mental representation of the object and its fate. At some level, in other words, they have a picture in their minds of the object and its trajectory through space, even when that trajectory is hidden from view.

The ways cats encode in their minds their perceptions about the world have been probed in other recent studies that involve searching for hidden objects. In one series of experiments, which could easily have been inspired by the Monty Python "confuse a cat" sketch, cats were subjected to all sorts of disorienting visual tricks between the time they saw a toy hidden behind one of several identical-looking screens and the time they were allowed to go and search for it. In one test the toy was placed behind the rightmost of three screens, then with the cat's view momentarily blocked, all the screens were slid over to the right by a distance exactly equal to the spacing between them. In a more elaborate version of this experiment, the cat was allowed to look into the experimental chamber from the doorway; and after the toy was hidden the entire room, walls and all, was shifted to the right a like amount while the cat's view was blocked. Quite consistently, when the cats were released and allowed to hunt for the toy, they would use an absolute sense of position rather than a relative one. That is, they did not look for the toy behind what was now the rightmost screen, but rather they looked behind the screen that now occupied the precise spot in space that the rightmost screen had previously occupied at the time the toy was hidden. Their sense of space was thus "egocentric"; the cats remembered where the object was placed relative to their own fixed position in space, and not by the object's position relative to some landmark.

Only when the experiment was set up in such a way that it was impossible to use egocentric spatial reasoning were the cats forced to use landmark orientation, and in that case they did so successfully. (This was a test in which the cats were forced to take a detour through an L-shaped tunnel and enter the room from a door to the left or right from the one through which they observed the toy being hidden.) But whenever the egocentric cues and the landmark cues conflicted, the cats trusted to their own cat-centered coordinate system.

Cats do, then, form a mental map of their environment, but it is an interesting sort of map, one with a cat in the middle. Some of the odd things that cats do in relating to the world, things that at first blush can seem awfully stupid, may be explained by this sort of spatial mental representation that cats favor. I have more than once moved the food bowl of one of my cats to a different spot—and he can be sitting right there and watch me do it—and he will then go directly to where the food bowl used to be and look around for it with an air of perplexity.

It is certainly tempting to believe that given such sophisticated abilities to remember experiences, to make complex decisions, and to form mental representations or perhaps actual images of objects in the world, cats do "think" more or less as we do. The trouble is that we cannot say for sure, and it is not just pedantic scientific skepticism to be cautious. Cats unquestionably do think in one sense; they take in information about the world, perform mental operations on that information, and make decisions that they then act upon. And they clearly *know* things: They know where food can be found and what time of day things happen; they know by sight and smell the individual identities of the people and other cats they regularly encounter; they know the bounds of their home territory and what a toy looks like.

But we know from many experiments, in both human beings and other animals, that it is possible for a being to know things without being aware that one knows them, and to think without being conscious of thoughts. Indeed, the mental tasks that I have just described cats performing are the kinds of things that human beings generally do without conscious thought or awareness at all. We can recognize a familiar face or find our way down a familiar street or estimate the passage of time without even being able to explain how we did it. Complex mental processes don't necessarily translate into the human sort of self-awareness that involves not just having thoughts, and not just knowing, but having thoughts about our thoughts and knowing that we know.

One clue that cats may lack a fully developed sense of self-awareness is that like dogs—but quite differently from some monkeys and apes—they flunk the so-called mirror self-recognition test. If you show a chimpanzee a mirror, he will use it to explore the inside of his mouth, for example, and do other things in front of the mirror that seem to indicate he understands he is seeing an image of himself. Cats react to their mirror image at first by treating it as a hostile intruder, and then rather quickly learning to ignore it altogether when it fails to react like a normal cat. (Sometimes they look behind the mirror as if hunting for the cat they saw in it.)

Teaching and Watching

Not long after kittens start to be able to move about on their own and leave the nest regularly, their mother changes the way she captures and eats prey. Instead of catching a mouse and eating it on the spot, she starts to carry it back to the nest and eats it front of the kittens, and may let them eat some of it. Next she typically brings back a live mouse and carries it directly to the kittens and releases it, allowing the kittens to play with it, but recapturing it if it tries to get away. Finally she will leave it all to the kittens.

Laboratory experiments have shown that just as mother cats seem to teach, so kittens seem to learn by watching their mothers. In one set of experiments, 9-to-10-week-old kittens who watched a strange adult "demonstrator" cat obtain a food reward from a test apparatus learned to perform the task themselves when allowed to use the training apparatus (the cats had to learn to press a lever when a flickering light was turned on); kittens who had not watched a demonstrator cat perform never learned to do it in the course of the 30-day experiment. Kittens whose mother served as the demonstrator learned even faster; they began pressing the lever in response to the light after 4½ days on average as opposed to 18 days, and it took them an average of just 3½ additional days to learn to press the lever consistently on cue, as opposed to 14 days for the kittens who had the strange demonstrator.

As impressive and intriguing as these findings are—for, besides cats, only primates have ever been shown to learn by observation in a similar fashion—there is a great dispute among animal behaviorists about just what is going on here. Skeptics point out that what the kittens may actually be learning by watching a demonstrator cat is simply not to be afraid of a novel situation or a strange new gizmo: Those kittens who saw the process were emboldened to approach the apparatus and poke around with it for themselves, but it was still up to them to do the actual learning on their own. Such a process is more properly called "facilitated learning," as opposed to "observational learning." Emotional and social attachments would make the reassurance more powerful when the kitten's mother was the one to press the lever.

Likewise, the learn-to-hunt lessons that mother cats provide their kittens may be important for habituating them to what might otherwise be at least a somewhat frightening or suspicion-arousing stimulus of a live mouse; the lessons are also important in providing the kittens a far greater opportunity than they would ever have on their own to exercise their instinctive hunting response, and so learn how to adjust or alter or fine-tune it to maximize its effectiveness.

A somewhat more liberal interpretation of these events is that by seeing another cat do it first, the kittens learn that it is possible to effect certain changes in the state of things in their environment; they may not in a single gestalt understand that pressing a lever when the light flashes yields food, or how to use their claws to grab a mouse, but they learn that there is *some* connection between pressing a lever and the provision of food, or between using the claws and successfully catching and holding a mouse, and so are motivated to discover the precise relationship by trying it for themselves. One experiment even found that cats can be misled by what they see other cats do: Cats who had first observed a demonstrator cat receive food when he walked up to, but did not press, a lever on an apparatus had a much harder time subsequently learning *to* press the lever for food than did cats who had not witnessed the misleading demo.

A true ability to see and imitate what another animal does, however, seems to be extremely rare in the animal world—and some scientists who have studied the problem carefully argue it is, so far as we can tell, nonexistent. Cats are probably no exception to this rule. Cats learn more readily when they are in a comfortable social situation; they learn more readily when an opportunity for learning is created; but they still have to try it themselves, and learn the consequences firsthand, for the lesson to stick.

SIX

The Cat Personality Test

Domestic cats vary enormously in their individual personalities, probably more so than any other species, domestic or wild.

Animal behaviorists in the past often discounted the significance of individual differences in behavior they found among members of the same species, just chalking it up to the incidental vagaries that always exist in living things. It was cats, however, that helped shake that opinion. The differences in cat personalities were just too great to ignore when scientists began doing behavioral experiments on lab cats, and quantitative studies of cat personality have since confirmed this. There are bold cats and shy cats, friendly cats and hostile cats, easy-going cats and nervous cats, and these differences are not in any way the result of abnormal behaviors or mental pathologies; perfectly normal cats all, they just come in many, very different flavors.

One of the reasons behaviorists used to think that all members of a species ought to act the same was the evolutionary argument that behaviors evolve for their fitness to the environment, and all members of a species live in the same environmental niche. But studies of individual personality have begun to show that often there is an advantage to being different. Doing something different from everybody else can often be a winning strategy—as was the case among the feral cats that Eugenia Natoli studied in Rome, where "unconventional" easygoing males would slip in and mate with a female in heat while the conventional macho males were busy fighting one another. Evolutionary pressures do not always select for uniformity.

Cats may be particularly plastic in their development of personality because of the natural social transformation that the species experiences during the transition from infancy to adulthood. Cats evolved in their natural state to go from group-dwelling social animals as infants to loners as adults. Just as domestic dogs have radiated into a wide range of adult physical forms as a result of small genetic perturbations in the rate and timing of the dramatic physical changes that take place as wolf cubs mature into wolf adults, so domestic cats may acquire such a range of adult personality types as a result of small social perturbations that occur in the course of the dramatic social transition that they are programmed to experience.

As a result, there appear to be a number of ways that cat owners and breeders can dramatically influence the personalities that the kittens they raise will have as adults. Some of a cat's propensity to be sociable and easy-going appears to be inherited, and inherited from the father in particular. But experimental studies of cats also clearly show that everything from learning ability to friendliness toward humans to playfulness is strongly determined by the experiences they have had in kittenhood.

The Cat Personality Test

Categorizing cat personalities is an inevitably subjective business. One way to make it more objective is to hold off on drawing any conclusions about overall personality types ("personable," "domineering," "shy," "affectionate," and so forth) and instead see whether data on the individual behavior patterns of a number of cats organizes itself naturally into certain patterns or clumps. This is what Julie Feaver and her colleagues at Cambridge University did. They started with a list of twenty-seven fairly objective categories of things a cat might do, such as sleep; sit; purr; approach another cat or a person; rub his head on a cat, person, or thing; roll on his back; run away from another cat; chase another cat. Human observers watched the fourteen cats in Feaver's Cambridge lab every day for three months, and for two 5-minute intervals each morning recorded at 15-second intervals exactly what each cat was doing, using the checklist of twenty-seven specific behaviors.

At the end of the three months—during which time the observers were told not to speak to one another about their observations, in order to make sure everyone's judgment was as independent and unbiased as possible—the observers rated each of the cats on various overall behavioral characteristics ("active," "curious," "fearful of people," "tense," "solitary"—eighteen in all). The observers were asked to record their ratings by marking an X on a scale that ran from – to +; that marking was then converted into a numerical score that was "normalized" so that the average fell at precisely 0 and most cats' scores were in the range from −1.0 to +1.0. A negative score thus meant that a cat was below average for the trait; a positive score meant above average. Then, the scientists analyzed these behavioral ratings with another statistical measure to see how closely the observers all agreed with one another, and threw out the categories on which it was impossible to find much of a consensus. Finally, they looked at what was left to see if any patterns emerged.

Most obvious was the fact that some behaviors were strongly correlated: They tracked together in all cats. For example, the researchers found that a cat might either have positive scores on both active behavior and curious behavior, or negative scores on both; accordingly the researchers combined these two scores into a single new general behavioral characteristic that they dubbed "alert" behavior. Likewise, cats who scored positive on sociable behavior toward people scored negative on fearfulness toward people, hostility toward people, and tenseness; and vice versa. Scores for these four behaviors were combined into the general category of "sociable" behavior. "Equable" behavior toward other cats was a third general behavioral category that emerged from this analysis of the raw data.

None of these three general categories showed an across-the-board correlation with any of the others, and thus they appeared to represent "reasonably independent dimensions" of a cat's personality.

But some striking correlations did appear within certain subgroups of the cats studied. That is, there were certain distinctive personality profiles that showed up repeatedly. There was one group of "bossy" cats who all scored highly positive on the alert and sociable categories and highly negative on the equable category. These cats often threatened other cats and rarely retreated. A second group of "timid" cats scored highly negative on all three of the overall categories. These cats were nervous and often tried to stay out of the way of the others. A third group, the "easy-going" cats, were strongly positive on all three categories. They were highly sociable, confident, and rarely seemed either to threaten or be threatened by other animals.

Probably few cat owners have the objectivity or patience to completely reproduce what Feaver and her co-workers did, and to sit for months on end recording what their cat is doing every 15 seconds for 10 minutes a day. But the basic formula does make for an interesting personality test that anyone can administer to see where their cat falls on this scale. For the rating system to work, you need to have a fairly good sense of the range of behaviors that different cats exhibit so that you can make a good judgment as to whether your own cat is above average, average, or below average. The following is an adaptation of Feaver's test that ought to give a reasonably accurate result.

First, for each of the behaviors described on the next page, score your cat 0 if he or she exhibits the behavior less often than the average cat does, 1 if about the same as the average cat, and 2 if more often than the average cat.

a. Active	Moves about frequently	____
b. Curious	Approaches and explores a change in the environment	____
c. Sociable with people	Initiates proximity and/or contact with people	____
d. Fearful of people	Retreats readily from people	____
e. Hostile to people	Reacts with a threat and/or causes harm if approached by people	____
f. Tense	Shows restraint in movement and posture	____
g. Equable with cats	Reacts to others evenly and calmly, not easily disturbed	____

The next step is to combine these scores into the three more-general ratings:

1. Add together the scores from lines a and b ____ = A (Alert)
2. Add together the scores from lines d, e, f ____
3. Write the score from line c ____
4. Add 6 to the amount on line 3 ____
5. Subtract line 2 from line 4 ____ = S (Sociable)
6. Write the score from line g ____ = E (Equable)

A score of 3 or higher on A, 5 or higher on S, and 0 on E fits the profile of the "bossy" cat personality.

A score of 1 or lower on A, 3 or lower on S, and 0 on E fits the profile of the "timid" cat personality.

A score of 3 or higher on A, 5 or higher on S, and 2 on E fits the profile of the "easy-going" cat personality.

The researchers who did the original study have cautioned

that their results may not be applicable to all cats, since this rating scheme was developed using only a group of cats living together in a laboratory colony. Nor does it take into account all the many nuances of cat behavior and personality.

PERSONALITY AND BEHAVIOR

Personality Type	Scores on General Behavior Categories		
	Alert	*Sociable*	*Equable*
Bossy	high	high	low
Timid	low	low	low
Easy-going	high	high	high

On the other hand, the living conditions of these lab cats were not that different from those of many pet cats; they lived in two groups, each of which had a 4 × 4 meter indoor pen connected to a 4 × 5 meter outdoor pen (13 × 13 feet and 13 × 16 feet), about the size of a two-room apartment. The cats all had many hours of contact each day with friendly human beings. And the range of behaviors the Cambridge researchers found in their lab cats certainly seems to take in a broad cross section of the social spectrum that can be found in any domestic cat population.

Emotion, Neurotransmitters, and the Father Factor

Cats of different personality types are clearly different emotionally; not only do they act differently but also they must be subjectively experiencing events in different ways. So personality differences are potentially a way to probe a bit deeper into the mental life of cats and get a glimpse of the emotional worlds they occupy.

The trouble is of course that emotion is by definition subjective. Certain objective manifestations of emotional arousal that can readily be measured—things like heart rate, blood pressure, respiration rate—can show when an animal *is* aroused but say next to nothing about what particular emotion is behind that arousal. In fact, in people as well as in other animals, emotions as diverse as fear, sexual excitement, and anticipation of play or food can be indistinguishable as far as the physiologic changes that accompany each.

A further complication is that given the human propensity to confuse the issue by thinking about things, our own categorization of emotions probably is a very poor guide to the way even our own brains actually experience and divvy up emotions at a basic level. We have a great many terms for some emotions and no clear terms at all for others. Depending on the social context in which it occurs and our own conscious understanding of its cause and significance, an emotion like sadness can be described as melancholy, anguish, grief, pity, loneliness, despair, disappointment, regret, and probably a few dozen other terms.

That said, there is a growing body of evidence from neurophysiological research that is beginning to show how the brains of animals process emotions, and also exactly why, on a basic level, individuals differ so much in their emotional responses. In cats, both fear and rage are mediated by complex electrical and chemical pathways that have many potential influences. But at least some of the differences between individuals appear to be a matter of inheritance: Cats who are more fearful or more aggressive have very different patterns of electrical wiring in the circuits of the brain that regulate emotional reaction, and perhaps also some differences in levels of brain chemicals such as dopamine.

A series of particularly remarkable studies found that when certain precise spots in the brain are stimulated electrically, cats will display a characteristic fearful or aggressive response. The spots that trigger such reactions are located in the relatively "primitive" parts of the brain where control of basic bodily functions are centered: the brain stem, and also a specialized organ of the forebrain called the hypothalamus. But both reactions can be greatly enhanced or reduced, or shut off altogether, by electrically stimulating other parts of the brain—particularly the amygdala, an almond-shaped mass of gray matter whose function is closely tied to mood, instinct, and feeling. Stimulating one part of the amygdala can make it easier to trigger defensive rage in the cat, either when the right spot on the hypothalamus is directly stimulated with an electrical impulse or when the cat is presented with a threatening stimulus, like a tape-recorded yowl of another cat. Stimulating a different part of the amygdala makes it easier to trigger a predatory attack, and harder to provoke defensive rage.

Detailed studies that have traced the way nerves interconnect in the brain have found that nerve cells extending from the amygdala project directly into the mid-brain, and there they can apparently act to either enhance or block the flow of commands sent out from the hypothalamus. Threatening stimuli provoke a much greater electrical response in the amygdalas of more fearful and defensive cats than they do in less defensive cats, apparently a reflection of some actual differences in neural wiring.

Studies have also found that when cats are injected with a chemical that acts like dopamine, it is easier to trigger predatory attack. Dopamine is one of many neurotransmitter chemicals in the brain; it is released by nerve cells and it acts as a "messenger" to stimulate other adjacent nerve cells to fire. An excess or deficiency of dopamine or other brain chemicals, including the natural opiates the brain itself produces, called enkephalins, may thus have a dramatic effect on an individual cat's propensity for certain emotional reactions to stimuli.

People who have raised colonies of laboratory cats have long noticed that there were striking differences in the ease with which kittens of different parents could be handled, even when the kittens were all raised under identical conditions. A few years ago several researchers made a careful study of this phenomenon and proved statistically that this anecdotal impression was undoubtedly true. In one study, Sandra McCune of Cambridge University separated kittens into two groups; all of the kittens remained with their mothers and littermates, but half were handled by a person daily for a total of 5 hours a week from age 2 weeks to 12 weeks, while the other half were exposed to people only in the course of daily cleaning and feeding. Each of the two groups included litters of kittens whose fathers were friendly toward people and those whose fathers were not. The fathers were not present while the kittens were being raised, so the only effect they could have had was through their genes.

When the kittens were a year old, they were all tested to see how they would react to meeting and being handled by people and how they would react to strange objects. The effects of early handling and the effects of having received the genes of a friendly father turned out to be about equal in this experiment. For example, handled cats with unfriendly fathers and unhandled cats with friendly fathers were each about as likely to hiss when a familiar person approached them; both were less likely to hiss than the unhandled, unfriendly-father cats.

It is not clear whether this inherited tendency is something more strongly influenced by the father than the mother. One study did report a maternal effect as well, but it is extremely hard to distinguish the effects of genes from the effect of learned behavior on the maternal side when the mother is left to bring up her litter.

The cats who were friendlier toward people in McCune's study were also bolder in approaching and investigating a strange object, which suggests that the major emotional difference between friendly and less friendly cats may be in their general level of fearfulness. The fact that researchers have been able to artificially manipulate a cat's fear response so strikingly in the lab through electrical stimulation of selected brain pathways or by altering levels of brain chemicals is at least a strong hint of how the natural tendency toward friendliness is controlled and inherited.

The Developmental Psychologist's Guide to Raising the Perfect Cat

In animals, just as in people, emotional reactivity and personality are affected not only by the brain one is born with but also by early experience in childhood, learning over time, and occasionally sudden shocks such as an abrupt change in environment or a traumatic experience. The emotion-regulating circuits of the cat's brain have individual variations in wiring that are inherited, but these circuits also can begin to behave differently as a result of experience. Cats who learned to anticipate a fearful stimulus that was always preceded by a tone developed a permanent change in the way the nerve cells in the amygdala fired. The cells fired more rapidly, and they also became more synchronized, firing together in a pulsating pattern that appeared to help coordinate interaction with areas of the more "thinking" part of the brain, the cortex, that are involved in memory. Experience and memory seem to actually change the way the emotional circuitry responds to stimuli, in a lasting way.

It is well established from a number of experiments that cats who are handled by and exposed to human beings early on in life are much friendlier toward people throughout their lives than those who have been isolated from people. In studies in which kittens were handled and petted over a course of several weeks beginning at different periods of life, those who received the handling from age 2 weeks or 3 weeks scored much higher than those whose handling sessions began at age 3, 4, or 7 weeks when they were tested later for their tolerance of being held and their willingness to approach a person. Kittens who had been handled from a very early age, beginning at 1 or 2 weeks, would walk across a room and approach a seated person in an average of 9 seconds from the time they entered a room, versus about 40 seconds for cats whose handling began at 7 weeks of age. Cats who were not handled at all in kittenhood took almost exactly the same time as the 7-week group. These studies have also shown that it seems to make little difference whether kittens were handled from, say, age 3 to 7 weeks or age 3 to 14 weeks. In other words, there seems to be a critical window from about 2 weeks of age to 7 weeks of age when kittens are particularly receptive to forming lifelong attachments to people. Compared to the prime socialization period for puppies, this period in kittens starts earlier and ends earlier, it also ends more abruptly.

It is important to note that it is emphatically not necessary that kittens be with their eventual, permanent human owner during this socialization period; all that matters is that they are handled by *some* human being during this time. Removing kittens from their mothers and littermates before 7 or 8 weeks of age has many unfavorable consequences for emotional, social, and intellectual development, and so it is a mistake to confuse this socialization period with the age at which kittens should be adopted.

This socialization period in kittens bears some resemblance to the so-called critical period in species that go through the process of "imprinting." Imprinting, which was described by the animal behaviorist Konrad Lorenz in his famous experiments with geese, is the process by which the newborns of many species who are able to zoom around on their own as soon as they are born seem to form instant and almost irreversible attachments to the first things they see moving. Geese and ducks that form hopeless attachments to dogs, tractors, or animal behaviorists have been the staple exhibits of this phenomenon. The evolutionary explanation for this phenomenon is that such precocial young could easily blunder off on their own, so some special and powerful learning mechanism is required so that they right away learn to stick with their mothers.

But this sort of strong imprinting occurs in the first few hours of life, and it is not exactly what occurs in cats. Kittens are born helpless and are not about to zoom anywhere. That gives them much more of a chance to learn via the normal processes of learning the facts of who their mother is and where home is. The attachments that kittens form during this period of socialization are not as powerful or irreversible as that which occurs in true imprinting. As a result, the usual term that animal behaviorists now apply to this period in the life of species such as cats is the "sensitive period," which is meant to sound rather less definitive than "critical period."

Even that may be a bit of an overstatement if it is meant to imply that some special mental processes switch on for a few weeks and then shut down. Rather, it may be that the existence of the sensitive period for socialization in cats is the net result of a number of largely social factors that are at work in the developing kitten's life.

One factor is that it is simply easier to learn things the first time than to unlearn them later, and kittens between the ages of 2 and 7 weeks are experiencing many firsts in their lives. Kittens usually open their eyes for the first time at 7 to 10 days after birth, but it takes 3 weeks before they are able to orient themselves using visual cues and to follow moving objects, and their vision still remains cloudy until about 5 weeks of age. Hearing becomes well developed at about 4 weeks. Kittens can barely move about on their own for the first 2 weeks of life; although rudimentary walking begins at about 2 to 3 weeks, and running at 4 to 5 weeks, it is only at 6 to 7 weeks that they start to use all of the gaits of an adult cat. So in other words, it is during this period from 2 to 7 weeks that kittens are first *able* to fully explore and perceive and discover things in their world. It is not surprising that this should be a period when strong initial impressions are formed.

A second factor that makes learning during this period important, and arguably unique, is that lifelong learning *patterns* may be set by what kittens learn in their first few weeks. There is ample evidence from experiments in cats (and many other species) that learning begets learning; young animals that enjoy a rich variety of visual and social and tactile experiences in infancy develop faster mentally and are more able to learn throughout their lives. Cats who lived free on a farm, for example, were found to be much more adept than laboratory-reared cats at mastering classical learning problems such as discriminating visual signals. Conversely, kittens deprived of social contact and stimulation may resist changing their ways later in life, even when presented with rewarding social experiences that other more supple-minded cats learn to adapt to.

And finally, kittens become naturally more fearful as they approach their third month of life, and that may be a serious barrier to learning to have friendly relations with human beings for cats who have not already had the chance to do so. Older cats learn not to be afraid of things only if they are willing—or forced—to hang around and find out that these things are nothing to be afraid of. Kittens under seven weeks are relatively fearless and so are more open to learning not to be afraid of things, like friendly people, that are in truth nothing to be afraid of.

A combination of emotional and intellectual learning seems to be involved in the socialization of kittens. Instinctive emotional patterns are altered by early experience, principally loss of fear of novelty. This is arguably "imprinting" of a kind; so, too, is the learning that takes place in kittens during this period of who is a cat and who is not. Adult cats who are confronted with a silhouette of a cat typically raise their hackles and approach it cautiously but show no reaction to a similar silhouette of a blob. Kittens only begin to show a similar response to a cat silhouette beginning around six weeks of age. Experiments in which finches of one species were cross-fostered with parents of another species show that these birds grow up confused about who is an appropriate mate; they will preferentially try to mate with members of their foster parent's species rather than their own. So some crucial wiring of instinctive impulses in the brain arguably does take place early on, when it comes to learning who is a member of one's own species. By injecting ourselves into the lives of cats during their first few weeks, we may thus stretch the developing kittens' notions of who is a cat to include human beings as at least honorary members.

But there's also just plain learning that goes on, through trial and error experience. Kittens learn, for example, to control their aggressiveness toward other cats through trial and error in playing with their littermates. At first they frequently make the mistake of biting too hard and suffer retaliation or the frustration of having the "game" end abruptly when their playmate runs off. (Kittens who are removed from their littermates and mother at an early age show much greater aggression toward other cats than do those who stay with the litter through the natural age of weaning. It is not easy to separate the emotional and developmental trauma of early separation from the loss of the chance to learn "intellectual" lessons about social behavior, but the latter is surely at least a factor.)

As I have mentioned in an earlier chapter, kittens also appear to acquire their adult social style as a result of what they directly learn from their interactions with other beings in kittenhood. Those kittens who live in a group of friendly adult cats simply have many and repeated opportunities to learn that interacting with other adult cats is a positive experience. The same is true about their interactions with people. At a stage of life when they are curious and unafraid, they learn a lesson they would shy away from even having a chance to learn at a later stage of life: that people are friendly and will pet them and play with them.

Given the facilitated learning that occurs in cats, the actions of a kitten's mother are also an instructive and molding influence. Kittens who were exposed to people with their mother present were at first less likely to interact with the person; their mother in effect competed with the people for the kittens' attention. But over time, these kittens more readily initiated contact with people than did kittens who had been first exposed to people without the mother in the room at the time. Apparently the reassuring presence of the kitten's mother and her calm behavior around people facilitated the kittens' subsequent willingness to approach a person.

The real take-home lesson is that kittens from friendly parents benefit two ways: from their genetic predisposition and from the example their mothers provide.

Likewise, the power of early experience plays a decisive part in the future hunting ability and food preferences of kittens, a factor that anyone looking for a good mouser or a kitten who is not a finicky eater might want to consider. Because kittens learn to hunt largely by the opportunities to play with dead and live prey provided by their mothers—and because their wariness or even fearfulness of strange prey animals can interfere with subsequent learning if they do not have this early experience at their curious and fearless stage of life—kittens' hunting abilities are largely "inherited" from their mothers. This inheritance is, like social preferences, one of "culture," not genes, however. Furthermore, studies have found that cats tend to become specialists in one kind of prey if their mothers have a like preference. Cats who were experienced at hunting birds were not as interested in, or skilled at, hunting mice; and vice versa.

Finicky eaters also seem to be made, not born. In a slightly bizarre but revealing experiment, mother cats were first trained to prefer banana slices to meat pellets. This was done by rewarding the mothers whenever they chose the bananas by delivering an electrical stimulation to a region of the hypothalamus that evokes a pleasurable response. After they were trained, their weanling kittens were allowed to accompany their mother whenever she was given a food choice session, and they were allowed to see what went on. The kittens were then removed from the mother and given a free choice of foods. Fifteen of the eighteen kittens continued on their own to eat bananas in preference to meat pellets. (Kittens in a control group wouldn't even touch the bananas.) Similar results were obtained with the kittens of mothers trained to eat mashed potatoes or jellied agar.

So indulging the preferences of a finicky mother may be a recipe for producing finicky kittens.

Catnip Explained

Not all instinctive behaviors have a purpose. In a thing as complicated as the mammalian brain and its associated system of sense organs and chemical messengers, and in a world full of life as complicated as Earth, coincidences are bound to occur, and the cat's response to catnip seems to be one of those coincidences.

Only about half of all cats show a reaction to catnip, and the reaction is not affected by experience or learning. Cats who do respond typically begin by first approaching and sniffing the catnip source, whether it is fresh or dried leaves of the catnip plant, a toy stuffed with leaves, or an object sprayed with catnip extract. They then display one or more characteristic reactions, in no particular order: chewing or biting at the catnip source, rubbing or rolling against it, batting it with the front paws, or holding it with the front paws while biting it and kicking at it with the hind legs. Even cats who do respond to catnip generally seem to become satiated after about five or ten minutes of playing with it, and lose interest and walk away.

Some of these behaviors are remarkably similar to those of a female cat in heat, especially the rolling and rubbing that females display during courtship and immediately after copulation. Thus one popular theory is that the active ingredient in catnip—a complex volatile oil dubbed nepetalactone, after the scientific name of the catnip plant, *Nepeta cataria*—mimics a pheromone found in tomcat urine. But there are several problems with this neat theory. Both male and female cats are equally likely to respond to catnip. Cats on a catnip high do not exhibit other behaviors characteristic of female courtship, such as raising the pelvis. And as Benjamin Hart noted in his study of the catnip reaction, many behaviors associated with catnip-sniffing have nothing at all to do with sexual behavior. The biting motions cats direct at a catnip toy are reminiscent of the way cats bite into a rodent they have just killed; it is, in other words, just eating behavior. The batting and kicking motions are identical to the motions cats go through when they have grabbed a mouse, and are also typical of play behavior in general. Rubbing and rolling, for that matter, are not unique to sexual behavior but are also involved in territorial scent marking, social contact, submission, and play.

In other words, what seems to be happening is that all sorts of typical, hard-wired behaviors of the cat are triggered. Hart also found that the vomeronasal organ, the special receptor that is sensitive to sexual pheromones, is not involved in the catnip reaction; cats who responded to catnip continued to do so after their vomeronasal organs were surgically removed. Removing the olfactory bulb, the set of nerves that provides the sense of smell, did, however, extinguish the catnip reaction, as did temporarily anesthetizing these nerves.

Studies in other cat species have found that lions, jaguars, and spotted leopards were sensitive to catnip but tigers, cougars, and bobcats were not. But not all of the "responder" species reacted to it in the same way, and there was no particular correlation between species that instinctively roll over in estrus (not all cats do) and species that roll over in response to catnip. That is further evidence to suggest that catnip is not triggering a particularly sexual response.

There probably is no evolutionary significance to the fact that the catnip plant manufactures a substance that affects cats in particular; many plants happen to produce complex oils and resins for their own purposes, especially protection against insects, that by coincidence have psychoactive properties on mammals. It is true that some plants have evolved the ability to generate pharmacologically potent products as part of their natural defenses against birds and mammals (some clovers, for instance, produce an analog of estrogen that inhibits estrus in grazing animals, which perhaps is an evolved active defense: Plants fight back with birth control). But it is hard to imagine that the catnip plant ever had much at stake in the behavior of cats.

Nor does it seem that cats derive any useful benefit from catnip. If anything, the evolutionary evidence suggests that cats may have evolved *not* to have a reaction to catnip. The catnip plant is native to North America and Eurasia; and the particular cat species that are most sensitive to it are the ones that never would have naturally encountered it, while the species that are immune to its effects, such as cougars and bobcats, evolved where it was present. It is tempting to speculate that the ancestral population of *Felis silvestris*, with a geographic range that spanned both catnip-present and catnip-free zones, faced a mix of selective pressures for either carrying or not carrying the catnip-immunity gene or genes. Thus the population that gave rise to the domestic cat contained a mix of susceptible and resistant individuals, a mixed heritage that has been passed on to subsequent generations of house cats.

SEVEN

Cats and Trouble

Many problems that people have with their pets are the result of the unnaturally stressful conditions that domestic animals sometimes face in their modern lives. Horses that are boxed up in stalls many hours a day without the chance to engage in normal horse behaviors like running and socializing may develop an array of "stable vices" that are very similar to obsessive-compulsive behaviors in mentally ill human beings: They will chew repeatedly or weave back and forth or even engage in acts of self-mutilation. Dogs left alone all day may develop destructive compulsions or become hysterical when their owners leave. Cats are adaptable enough to a wide range of physical environments and social milieus that they are relatively immune to such mental traumas. Once they stake out their home territory, most cats are perfectly content to be left at home on their own all day, or even for days on end, even when home is a small apartment. And most cats who have been socialized in infancy readily tolerate the comings and goings and varied attentions of human beings, familiar and strange alike.

The trouble that people have with cats is more a matter of normal cat behavior than abnormal cat behavior. Cats don't go bonkers when kept in an unnatural situation. On the other hand, they don't particularly change their natural ways, either. And inevitably, some natural habits are incompatible with the ideas that cat owners have about the way they would like to run their lives, which often includes such things as owning a sofa without claw marks or a microwave oven that does not smell of cat urine. Given the resistance that cats put up to the direct kind of training that works almost automatically on social animals like dogs, these culture clashes between cats and people can become a major source of trouble and can often seem intractable. Some 4 million cats are euthanized each year in animal shelters in the United States, many of them having been relinquished as a result of behavioral problems. About half of cat owners surveyed reported at least one behavioral problem with their pet, and a review of the problems that drove particularly desperate cat owners to seek the professional assistance of animal behavior experts at Cornell University—generally a last resort after regular veterinary advice and treatment had failed—found that 59 percent were problems of "inappropriate elimination," 25 percent were problems of aggression toward either other cats or people, and the remaining 16 percent were a grab bag including eating problems (such as anorexia, wool eating, and eating houseplants), excessive vocalization, scratching, and various "activity problems."

Successfully dealing with these cat problems is often more a matter of what might be termed "ecological psychology" than training or therapy. Because these problems are usually the product of natural instincts rather than of abnormal behavior, the solution is not so much to "cure" the cat of a "disorder" as it is to find ways to channel and redirect the cat's natural instincts in a manner that the people in his life can tolerate. It's more a matter of tricking the cat into believing that the desired outcome is what he himself really wanted to do all along.

Messy Cats

One of the major selling points of cats is their neat elimination habits, which obviates the need for house-training and frequent daily walks. While it is true that cats have a natural instinct to defecate and urinate in litter, especially when in or near what they perceive to be the core of their home territory, that instinct is directly at odds with an equally powerful instinct to use urine, and sometimes feces as well, as a means of communication. Spraying urine on prominent vertical surfaces like trees and rocks, especially if they bear odor marks of other cats, and leaving piles of scats ("middens") along frequently used trails are well-observed behaviors in feral cats; and houses and apartments provide an abundance of analogous features. Cats have been reported to spray urine on walls, furniture, large appliances, stereo speakers, videocassette recorders, and countertop microwave ovens; in one case a cat started a fire by spraying an electrical outlet. Indoor cats who defecate outside of their litter boxes usually choose prominent areas of traffic such as doorways and halls that are analogous to trails and trail junctions. About 12 percent of male cats and 4 percent of female cats are "habitual" sprayers of urine around the house; as many as 30 percent will spray sporadically.

Many things can trigger the urge to spray, but at the root of this behavior is the cat's territorial instinct. Anything that provokes this instinct increases the odds of spraying. Most commonly, the cat is responding to a perceived intrusion on its territory. In fact, the incidence of indoor spraying is almost directly proportional to the number of cats in a household. In households with ten or more cats, the odds that some urine spraying will occur approaches 100 percent. Introducing a new cat is very often the trigger that sets off a bout of spraying in cats who do not habitually do so. Installing a cat door is another frequent trigger, especially if a strange cat happens to come through the door one time. The resident cat can feel territorially challenged after even a single incident of this sort, and continue spraying for days afterward.

Interestingly, though, it's often the newly introduced cat rather than the established cat who is the one doing the spraying and middening in a multicat household, and the explanation for this involves a combination of cat psychology and human sociology. Often the person who gets a second cat does so because the first cat has proved a bit disappointing—typically because he is timid and unfriendly. The new cat has thus been selected precisely because he is, by contrast, outgoing and active. But outgoing and active cats are also "bossy" and dominant cats, and dominant cats are more likely to spray and midden when confronted with a territorial challenge.

Resident cats who suddenly begin spraying may have been set off by intrusions that humans are not even aware of but which loom very large in the cat's view of the universe. Cats may begin spraying in response to just seeing a strange cat out a window; this is especially likely to be the explanation when the resident cat routinely sprays at windows or doors. (Likewise, the arrival of a new baby sometimes triggers an upsurge in marking.) Even a strange smell can elicit a marking reaction. A new piece of furniture arriving in the house is often a target for this reason. (My barn cats never pass up an opportunity to spray on new bales of hay that arrive from another farm.) Sometimes rearranging the furniture can provoke a spate of spraying, apparently as the cat perceives that there is a new prominent "landmark" in his territory which he feels he had better mark.

Because pet cats have come to view human beings as honorary members of cat society, human odors can carry as much weight as cat odors in triggering the urge to mark, which may be why household appliances that are frequently touched by people often become targets. Sometimes when an owner has been away for a while, a personal object like a suitcase or a coat that he plunks down on his return—and which carries a fresh and strong odor intruding into the cat's territory—becomes an immediate target.

Owners often ascribe cognitive explanations to such behavior, thinking that their cat is mad at them for leaving her alone and is exacting her revenge, or is jealous of the new baby in the family. But the cat's basic marking instinct really explains completely what's going on in such cases. The cat isn't mad; she's just responding to something new and sudden in her territorial landscape.

Cats vary in their innate tendency to react territorially to such intrusions, and there is no doubt that castration and spaying greatly reduce the likelihood of spraying. Territoriality is very closely wrapped up with sex in most species, and cats are no exception. About 80 percent of male cats with a spraying problem show a very rapid decline in this behavior after castration. Because female cats tend to increase their marking when in heat, spaying likewise reduces the tendency toward spraying in females. Even among castrated males, however, the presence of a female cat in the household tends to bring out the territorial imperative; in one study by Benjamin Hart, castrated males with female housemates were twice as likely to spray (40 percent) as were castrated males with male-only housemates (20 percent). Apparently when there are no females in the picture, the males figure that there is not anything worth making a territorial issue over. (Females, for their part, were just about equally likely to spray regardless of whether they had male or female housemates.)

So one simple preventive measure is to castrate and spay one's cats and to have cats of the same sex only. Creating fewer opportunities for conflicts over territory and resources among the resident cats in a multicat household can also sometimes produce dramatic improvements, as can avoiding any sudden changes in the environment of a cat who is prone to react to novelty by increased territorial marking. A new cat can be introduced by confining him to a separate room with his own food, water, and litter box at first and slowly introducing him. Multiple food and water bowls and litter boxes—one per cat—can reduce conflicts, especially if they are well spread out. Keeping curtains closed or shutting cats out of rooms that afford a good view of neighborhood cats can help if the sight of strange cats is provoking the marking behavior.

Cats are remarkably adept at learning associations with anxiety-provoking events, and extremely resistant to unlearning them. For example, some cats who have been alarmed by a strange cat coming through a newly installed cat door come to associate the door itself with the trouble. Covering up the door may cure the problem, only to have it recur immediately as soon as the door is uncovered. And cats can remain in a state of arousal for hours or even days, which is another formidable barrier to unlearning such associations. So trying to actively train a cat to overcome his fear sometimes carries with it a bit of the feeling of trying to reason with a hysterical person. Cats can sometimes be successfully trained to become inured to the fear- or anxiety-provoking stimuli that triggered urine spraying if they can be gradually reintroduced to them over time, but it's often an uphill battle with cats who already have a strong tendency to react to things they don't like with fear, aggression, or flight. Because cats don't learn very well by punishment, either, and because spraying is such a basic instinct, yelling at a cat or hitting him for spraying is pretty much futile, and can even be counterproductive if the cat interprets it as a ratcheting up of the social and territorial conflict with a rival "cat."

Cats who tend to spray in just one selected spot or who dump scats as territorial markers on a favorite place (like the top of a refrigerator, in the case of one particularly interesting perpetrator) can *sometimes* be conditioned by classical learning-theory methods to knock it off. One approach is to place upside-down, loaded mousetraps in the places the cat tends to use. When the cat touches the trap, it springs up in the air with a satisfyingly startling effect. This kind of "remote" punishment avoids the problems of turning punishment into an aggressive, social challenge; cats tend to view it as a mysterious and inexplicable phenomenon of nature, rather than a social interaction. (Cats sometimes can be taught by direct, "interactive" punishment not to perform certain behaviors, but the rule they usually learn from such experiences is an extremely "local" one: Don't spray when this particular person is standing nearby.)

Other forms of "remote" punishment for which some success has been reported include the use of a powerful squirt gun, an aerosol spray of something with an aversive odor like underarm deodorant, or a loud noisemaker such as an air-powered horn. The trouble is that the person not only has to catch the cat in the act to use such punishments but also has to activate the device of choice in such a way that the cat doesn't associate it with the person. Worse, some cats apparently come to savor the act of dodging a spray of water as a fun game, which can rather defeat the purpose. And some reports indicate that cats are smart enough to associate even remote punishment with the booby-trap itself rather than the place where the punishment takes place; once the mousetraps are removed, the cat goes right back to his old habits.

The ecological engineering approach probably works better even for these cases where a cat is marking in selected spots. Cats don't like to urinate or defecate near where they eat, so sometimes all it takes is just moving the cat's food bowl to the affected area to make the cat stop.

Some particularly tough cases—the cat who began to regularly defecate on his owner's bed certainly would qualify in this category—can sometimes be treated effectively with the help of psychotropic drugs. The most successful long-term results from the use of drugs reported in the scientific literature seems to be in cases where the drugs serve as a stopgap for a month or two so that permanent behavioral modification can have a chance to work. The idea is to reduce the cat's territorial instinct or his overall anxiety level while his environment is slowly modified (for example, by confining the cat to one familiar room and then gradually allowing him free run of more of the house) or while the cat simply has the chance to become habituated to the frightening things that had triggered a reaction. Hormonal drugs known as progestins, which block the effect of male hormones, have been found effective in reducing spraying in males; antianxiety drugs such as buspirone, which stops or markedly reduces spraying in about 55 percent of cases, work via a different biochemical pathway, apparently by blocking neural transmitters involved in a variety of species-specific behaviors.

One complication in treating elimination problems in cats is that a great many other factors than simply territorial marking may be going on. There are a variety of behavioral syndromes cats can acquire that lead them to routinely eliminate outside the litter box but that really have nothing to do with territorial marking behavior. (There are also a number of physical illnesses that can cause loss of bladder and bowel control that may need to be ruled out.) When cats spray as a territorial mark they generally do so on a vertical object, and so sometimes it is possible to distinguish a spray from an ordinary elimination simply by seeing where and how the cat has directed its urine. Likewise, cats who defecate outside the litter box without any intention of making a territorial announcement of the fact often do so in hidden spots rather than the prominent locations associated with middening.

In the case of such nonterritorial out-of-the-litter-box elimination, some truly bizarre learned associations are often found to be the ultimate cause. In a review of the surprisingly vast scientific literature on such "feline inappropriate elimination" syndromes, the veterinarian Leslie Larson Cooper found a remarkable variety of different phenomena at work. In multicat households a more dominant cat may view the litter box as part of his particular core territory and intimidate other cats who try to use it. Sometimes cats develop an aversion to a litter box because of its location, especially if it is placed close to their food or water bowl. Sometimes they develop an aversion because of some unpleasant event that happened there. Owners who have to give a cat medications sometimes find it easy to grab the cat while he is occupied in the litter box, and these cats are prime candidates for associating the box with Something Bad. Cats who have urinary obstructions or constipation that causes painful elimination likewise appear to readily associate the pain with the place, and often start avoiding the box. This is a particularly common syndrome in older cats.

Sometimes cats simply dislike certain kinds of litter or dislike using litter that is not changed often enough. Cats do develop personal, learned preferences for certain kinds of litter; and a cat who gets in the habit of liking, say, carpet, can be real trouble. The treatment in that case is to put a piece of carpet in the litter box—and confine the cat so he can't get to his traditional target areas—and then to gradually add litter over the carpet while snipping away pieces of the carpet. Cats who seem to dislike using the box because of the litter can similarly be confined with a "litter cafeteria" that offers an array of litter types for the cat to choose from. (One research study found that, a priori, cats do not on average seem to prefer one type of litter to another, with the exception of a slight bias in favor of clumping-type litters.)

Destructive Cats

Although elimination problems top the list of behavioral problems that drive owners to seek professional help, scratching furniture is probably a much more frequent problem among cats in general. One survey of cat owners found that 42 percent reported problems with scratching. Scratching, like spraying, is a fundamentally instinctive territorial behavior and is similarly dealt with most effectively through subterfuge rather than by classical conditioning. Feral cats and wildcats typically choose a specific, prominent, vertical object in their environment, most often a tree, to work over with their claws. The scratches themselves are a visual mark, and that is supplemented by the odor mark left by the secretions from glands in their feet. An indoor cat quite naturally looks upon something like the side of a sofa as a prominent vertical landmark, especially as it often has an attractively rough texture and stands in the middle of a room.

The difficulty in breaking the habit is that cats like to stick with the personal territorial marker they have chosen. The presence of their odor reinforces their tendency to return to it. Starting kittens with a commercial scratching post before they get any ideas is the best way to avoid trouble. A cat who has already begun to scratch the sofa can be eased out of the habit by pushing the sofa out of the way and protecting it with a cover and substituting a scratching post in the exact spot the cat has been scratching. The post can then be ever so slowly shoved to the side of the room, a few inches a day, and the sofa finally restored to its original position. Benjamin Hart found that cats prefer to scratch materials with long, straight fibers that make it easy for them to drag their claws through; nubby, tightly woven fibers are less attractive. Again, punishment is largely ineffective since the cat associates the punishment with the person rather than the place or his own action.

The second most common destructive behavior that cat owners complain about is eating houseplants (36 percent of owners). There is still debate over exactly why cats like to eat plants; but studies of feral cats have noted that they will eat grass nearly every day, so there is little doubt that this is a natural instinct. Booby-trapping the area around the plant can be effective, especially if the cat is at the same time given an acceptable alternative such as a small "cat garden" of grass. Benjamin Hart has also reported some success with a form of "aversion therapy," in which the cat is sprayed with aerosol deodorant a few times when caught in the act. The cat thereby learns to associate the smell of the particular brand of spray with an unpleasant sensation—most antiperspirant deodorants contain an active ingredient that is slightly irritating to the nose and eyes. Then the plant can be sprayed with the same deodorant so that the learned aversion is transferred to the plant itself. Hart notes that unless you wish to simultaneously train your cat to have an aversion to you, you should use a brand of deodorant different from the one you yourself use.

Vicious Cats

Aggression, like urine spraying, is a normal feline reaction to an intrusion in its territory. Cats who have been neutered and cats who as kittens have been socialized to people and other cats are more likely to tolerate group living and to react to a new cat or a new person in the house in an easy going way. The same procedures that can help reduce urine spraying when a new cat is brought in—keeping the new cat separated and confined at first, and introducing him slowly and allowing the cats to avoid one another as they wish by providing multiple food and water bowls, litter boxes, and resting places—can help prevent aggressive encounters.

But even cats who get along well with other cats and people sometimes begin to behave in aggressive ways that catch their owners by surprise. An otherwise good-natured cat may suddenly start stalking and leaping on and biting his owner; cats who have gotten along fine may begin hissing and swatting at each other every time they meet; a cat may repeatedly climb onto a person's lap and purr contentedly while being petted, then suddenly sink his teeth in.

Cats bite people far less frequently than dogs do, but cat bites are still a significant enough medical phenomenon to attract the attention of the kind of people who collect accident statistics. An epidemiological study in El Paso, Texas, found that the ratio of cat bites to dog bites was about 1 to 6; extrapolating that ratio to the entire United States suggests that about three-quarters of a million people are bitten by cats each year in America. The El Paso study also found that virtually all cat bites fell into the category that epidemiologists call "provoked." That doesn't necessarily mean that the person who was bitten was at fault, but it does mean that the person was doing something to the cat—picking him up, petting him—just before being bitten. The medical authorities find this data interesting in part because as many as half of all cat bites become infected, even though most of the injuries themselves are relatively minor and almost all cat bites occur on the hands or arms. On the other hand, one occasionally runs across a case such as that reported in Trois-Rivières, Quebec, in which an elderly man lost a pint of blood and required multiple stitches after his pet cat, Touti, went into a frenzy when he was accidentally hit by a spray of water while the man was, according to wire service reports, "giving his pet parrot a shower." An animal control officer who arrived to help found a scene of total carnage. "There was blood all over the place—on the ceiling, the floor, and the walls," he told reporters. (Unfortunately, the wire story added, "It is not known why" the man was giving his parrot a shower.)

Most cat bite cases arise from far less provocation than being hit with a stream of water. The cat who bites the hand that pets him is a phenomenon that has long baffled and upset cat owners, and one study found this to be "the most frequent form of aggression of cats toward people." Although it often seems that the cat bites without warning in these cases, there usually is some preliminary indication that the cat is becoming irritated; he may switch his tail or growl or make a sort of fake biting motion with his mouth toward the person's hand. What usually happens is that the person petting the cat ignores this warning, and finally the cat lets him have it for real.

There are various theories about what is going on in the cat's mind at times like this, but basically it seems to come down to the simple fact that cats don't like to be petted for as long as people like to pet them. When cats groom each other they tend to do it for short bouts of just a few minutes and the session ends when one gets up and walks away. A cat sitting on his owner's lap and being petted may feel a conflict between the pleasurable sensation of being petted and the unpleasant sensation of having his personal space violated. It also seems that some cats, at least, have only a certain level of tolerance for tactile stimulation and after being petted for a minute or two actually find the petting itself unpleasant. So even though a cat may initially "ask" to be petted, he can quickly become annoyed if it goes on longer than he wants it to.

The problem can be dealt with with virtually complete success by limiting petting sessions to short periods and watching for the warning signs of an annoyed cat; as soon as the cat shows signs of irritation the person can stand up and end the session.

Some cats may from time to time bite as soon as they are petted. These cases can be dealt with by withholding all petting for a few days, and then giving the cat very brief pets only when the cat initiates it, gradually increasing the petting sessions as the cat becomes desensitized.

Another common cause of seemingly mysterious attacks on people by cats is the syndrome sometimes referred to as "play aggression." Part of the normal repertoire of social play behavior in cats is stalking, grabbing, and wrestling. Most cats need a certain amount of play every day, and cats who don't have the chance to work out their urges by actual hunting or by play with another cat or by having someone pull a toy mouse on a string across the floor are prone to create their own play opportunities. Cats who stalk a passing ankle, or hide beneath a chair and spring out in ambush when a person walks past, sometimes are perfectly harmless and cute about it; but sometimes things can escalate into serious business if the cat starts using his teeth and claws. Kittens usually learn to inhibit their biting and scratching when playing with their littermates or friendly adults because if things get too rough their play partner abruptly loses interest and moves away, or sometimes retaliates. Some aggressively playful cats just never really learned this lesson and can be brought into line with punishment when things get too rough, such as a healthy squirt from a water pistol. But stronger or direct punishment can backfire if it makes the cat fearful, which may intensify aggressiveness under these circumstances. Some cats learn to stage highly effective hit-and-run attacks in response to attempts at retaliation, or they learn to attack when the person is particularly defenseless—such as by going after the person's hands and feet when he moves them under the bed sheets while sleeping. A more effective cure may be simply to make sure the cat has a few minutes a day of acceptable interactive play and to end any play sessions abruptly, for example by going to another room and shutting the door, if the cat bites.

Some people make the mistake of trying to "calm down" an agitated cat who gets carried away and bites, and they pet or hold or talk in a soothing voice to the cat. But the cat may interpret this as a reward or reinforcement for what he just did, and this can actually make the problem even worse. It may be a bit anthropomorphic to talk about punishing a cat with a "time-out," as some cat trainers advise, but, as a practical matter, isolating oneself from a cat who misbehaves this way is hitting him where it hurts: The cat learns that his attempt to initiate play by ambushing or biting simply doesn't pay off.

Because cats in a house or apartment cannot always readily flee from an anxiety-provoking social situation, they sometimes acquire rather weird learned associations between certain members of the household and a fearful or aggressive response on their parts. In one case, a cat happened to enter a room while another cat was hissing, having just been frightened by a rock hitting the window. The first cat apparently interpreted this as an aggressive display directed at him, and responded aggressively himself; the frightened cat responded more aggressively in turn. The two cats, who had always gotten along fine with each other before this incident, then began hissing and fighting whenever they met.

A variation on this theme is "redirected aggression," which happens in many species; it is best explained as simply a welling-up of aggression that seeks an outlet on some handy target. One study of redirected aggression in cats found that most cases were triggered when a cat caught sight of a strange cat through a window and then turned around and attacked his owner or another cat in the household. Cats may be particularly prone to such redirection because of their natural and strong reaction to territorial intrusion and because when aggressively aroused they often remain in a heightened state of agitation for a very long time—anywhere from half an hour to a day or more. A cat who is already on edge from having seen a strange cat out the window may react to small things like being picked up, which would otherwise not bother him in the least, and lash out.

And a cat who attacks his owner in such a redirected fashion may become conditioned to being fearful or aggressive toward that person. This is a textbook example of Pavlovian conditioning; in place of the thing that originally triggered the aggressive response, the owner who happened to be present now becomes the trigger. These cases can sometimes be very hard to diagnose because it is not always obvious what put the cat on edge to begin with or what particular coincidence caused the cat to associate a certain person with a certain frightening experience. These cases can certainly be very odd, and a unifying theme in all of them is the impossibility of anticipating or predicting the thing that first set the cat off. In one case reported in the scientific literature, a cat leapt on a child and bit her in the face the first time her talking doll spoke. Another cat began attacking his recently married male owner whenever the owner tried to clean out the cat's litter box. The attacks were quite serious; the cat lunged at the man's arms when he reached toward the box and scratched and bit violently. Although it was the man's new wife who logically might have been seen by the cat as the intruder, and thus been the target of aggression, in fact what had apparently happened was that the wife's presence had generally heightened the cat's level of social stress. The cat viewed the litter box as an area of particular territorial importance, and when the man moved toward it the cat found an outlet for his heightened level of territorial aggression. The man then became the Pavlovian trigger for this reaction.

Animal behavior experts report some success in treating such cats through gradual desensitization and "counterconditioning." For example, if a cat has suddenly learned to become fearful or hostile to one member of the household, someone else can feed the cat while that person stands on the far side of the room. The idea of counterconditioning is to have an animal do something—in this example, eating—that is incompatible with the behavior one is trying to eliminate—in this case showing fear of or hostility to a certain person. The feared person can move closer to the cat each day as long as his presence doesn't provoke the cat. In time, a new conditioned, Pavlovian association is thus learned to replace the old one.

Annoying Cats

Just as spraying the living room walls with urine is not really an "abnormal" behavior in cats—as incompatible as it may be with most people's idea of civilized life—many of the other annoying things cats are famous for really are just normal behaviors of cats with perfectly normal brains.

Cats are probably best classified as "crepuscular" animals; they are naturally most active around dawn and dusk, rather than being strictly nocturnal or diurnal. Cats do readily adapt to a wide variety of different schedules. But if left to their own devices, and if inadvertently rewarded by their owners for their behavior, they often will tend to do things like wake people up at 4:45 A.M. and demand to be played with one way or another. Pushing the cat away, chasing him around the room, or other such reactions may, from the cat's point of view, be perfectly satisfying forms of play and encourage him to do the same thing the next morning. Some cats tend to have a similar period of evening craziness, in which they run around a room and engage in all sorts of annoying attention-getting actions.

Other annoying but perfectly normal and natural cat behaviors that rank high on the lists of complaints from owners are jumping on tables or kitchen counters and excessive meowing. Indirect punishment via booby traps or squirt guns can be effective for these problems. One suggestion is to place double-sided tape on surfaces you want to keep the cat off. Cats find this surface unappealing, and it has the added advantage that cats have a hard time seeing the tape, so they tend to learn a "general" rule rather than a "local" rule—that is, they tend to avoid the place even after the tape has been removed.

Incessantly meowing cats are generally a much tougher proposition. Indirect punishment can be a problem, as some cats who get hit with water from a squirt gun when they meow learn to hide around the corner and meow.

In general, however, problems of excessive activity or waking people in the middle of the night or attention-getting behavior can often be helped by making sure the cat does get some playtime every day, especially before bedtime. If they get it out of their systems that may be enough to solve the problem. An analysis of how cats use their time shows that they spend about 85 percent of a typical twenty-four-hour period sleeping or resting. They do not in fact require a great deal of activity, but they do require some, and if not provided the opportunity they will generally manufacture it themselves.

Cats, as I have mentioned, are more resistant to the sort of

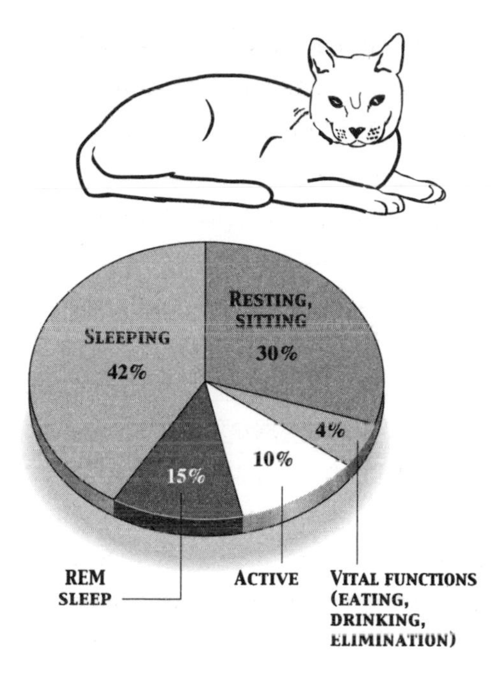

Cats spend about 85 percent of their time sleeping or resting.

truly abnormal, stress-induced behaviors that appear with a small but regular frequency in horses, dogs, and many wild animals kept in captivity. There is a whole class of abnormal behaviors known as stereotypies, in which an animal will repeatedly perform some natural action, such as walking or grooming or chewing or drinking water to the point of harm. These behaviors are strikingly similar to obsessive-compulsive disorder in humans. Studies in rats and other animals have shown that while initially provoked by stress, these compulsive behaviors appear to be sustained and reinforced by chemical pathways in the brain. Activities such as grooming tend to counteract the effects of stress by releasing endorphins, the natural opioid compounds in the brain. But that in turn can make these behaviors addictive, just as opioid compounds themselves are. Striking success has been found with cats who excessively groom themselves—in serious cases they do it to the point that they lose their fur and develop open wounds—by using drugs that block the effects of endorphins in the brain, breaking this cycle of addiction. Chemicals that block dopamine also have proved successful. Dopamine is involved in transmitting the nerve impulses for behaviors that are typical motor patterns in a species, activities like walking and grooming and eating. It also appears that stress reduces the threshold level of dopamine needed to transmit these signals. Thus chemicals that work against the effect of dopamine raise this same threshold, making it less likely that these species-specific behaviors will be triggered.

A few exceptionally weird cases of abnormal behavior in cats involve a subtle combination of such abnormal, compulsive behaviors and an inadvertent learned reinforcement by the cat's owner. Probably the world's record for odd cat behavior goes to the male cat who, in addition to urine spraying, would masturbate several times a week with a toy stuffed animal. The cat would mount the toy and grasp it on the neck just as males normally do while mating, and his owner reported further that the cat would be become extremely "agitated" if he could not find his toy, searching relentlessly for it while loudly meowing. The woman who owned the cat refused, however, to throw out the toy and it eventually became clear that in some way the cat was successfully using this behavior as an attention-getting device. His owner unavoidably paid attention to him when he went through this routine and she apparently even found it amusing at a certain level. Cats who are ignored and want to play learn all sorts of ways of attracting attention. (This cat had also learned, as other cats reportedly have, that spraying his owner directly with urine could be an effective attention-getting behavior.) Over time, a certain amount of Pavlovian conditioning had apparently occurred; the owner's presence had become the trigger that stimulated the cat's abnormal sexual behavior, and eventually the owner's return home after being away for the day was all that was needed to set the cat off. Although this cat probably had a hormonal and neurophysiologic tendency toward compulsive behavior to begin with, his owner's actions reinforced it.

Such behaviors are certainly not common, and few cat owners have to spend much time worrying that their cat will turn into a urine-spraying, toe-biting, fur-licking sex maniac. On the other hand there is arguably a useful warning lesson here about what things not to encourage.

Indoor Cats and Outdoor Cats

There is an odd bifurcation in the humane movement these days over the issue of indoor cats versus outdoor cats. On the one hand there has been a huge propaganda campaign on the part of many humane societies and self-appointed guardians of cat welfare to stigmatize cat owners who would even think of letting their cats venture out-of-doors. Some animal shelters actually refuse to allow cats to be adopted by people who will allow them outside at all. The "cats indoors" movement emphasizes not only the risk to cats of being hit by cars, injured in fights, and acquiring diseases such as feline leukemia from other cats, but also the danger to people that outdoor cats can pose when they pick up diseases such as toxoplasmosis from eating wild rodents. Toxoplasmosis is caused by a parasite that is easily spread to human beings through contact with cat feces in garden soil or litter boxes. In one case, thirty-seven patrons of a riding stable became infected apparently by inhaling toxoplasma cysts kicked up into the air when horses trotted around an indoor arena where the resident cats frequently defecated. Toxoplasmosis can cause birth defects, including blindness and mental retardation, when pregnant women are exposed to the parasite. Outdoor cats who hunt rodents are also a significant vector for transmission of plague to human beings in the American Southwest. Free-roaming cats are also a major reservoir of the rabies virus in America.

But probably the most prominent, and most effective, argument advanced by the "cats indoors" movement has been that free-roaming cats have a devastating impact on songbirds and other wildlife. Estimates that put the toll at upward of a billion birds and five billion rodents killed by cats in the United States annually are frequently cited by humane organizations and bird conservation groups in their campaign to get cat owners to keep their pets shut in.

At the same time, however, there has been a growing and increasingly zealous movement on the part of some cat advocates to capture, neuter, and re-release wild colonies of feral cats, which they maintain and feed. The advocates of the trap-alter-release programs insist that this is a humane alternative to rounding up and euthanizing stray cats that reach problem proportions and that have no hope of being adopted as pets, given their wild and ornery ways. The idea is that in time, these colonies will die out through natural attrition. The trap-alter-release advocates also vehemently (indeed some might say religiously) deny that maintaining large colonies of feral cats poses a significant threat to wildlife.

It is true that many of the frequently cited statistics of bird mortality due to cats are probably unreliable. They are extrapolations of limited samples, and there are many unknowns that affect the global picture. Individual cats vary enormously in the number of animals they care to kill even when allowed to roam freely; for some it is over a thousand a year, but for many it is zero. It is also very difficult to know how many of the animals that cats kill would have been killed in the absence of the cats by other predators, so it is at least possible to argue that cats, on balance, have little net impact on the overall population of even endangered and threatened species they prey on. (On the other hand, to the extent that cats do just substitute for the effect of other predators in this manner, they clearly are encroaching on these other predators' food supplies; and so even when they take abundant and common species such as field mice and house sparrows, they may be indirectly harming other rare and threatened species such as hawks and owls.)

But on balance, the evidence is hard to deny that cats have at least the potential to cause devastating ecological effects. Feral cats have been shown to be directly responsible for the extinction of eight island bird species on islands off New Zealand. A study in Britain reported that 30 percent of birds found dead had been killed by cats. A study of two California parks with identical bird habitats found that in one park, where more than twenty cats in a feral colony were fed by people dedicated to their protection, no California quail or California thrashers could be found, even though these species were present in the other park, which had no cats. Critics also point out that given the steady influx of new stray and abandoned cats to these managed and fed colonies, it is doubtful that natural attrition will cause them to vanish any time soon. Many conservationists believe that the fed and managed colonies of feral cats actually attract additional cats who may evade being trapped and neutered, and they also become "dumping grounds" for people who abandon their pet cats. And even if outdoor cats do not cause extinctions or other irreversible impacts on biodiversity, they certainly cause much pain and suffering to the billions of individual animals they kill. That fact alone poses something of an ethical challenge to the humane justification for maintaining and feeding large colonies of feral cats.

What has not been very well addressed in all of these often vehement debates over the effect of cats on wildlife is what effect keeping cats indoors all the time has on their own psychological welfare. No systematic studies have been done on this point. It does seem that many cats adapt without difficulty to a perpetual indoor existence as long as they receive a certain amount of daily stimulation and the chance to run and play. On the other hand, there are clearly some pet cats who can never adapt to indoor life. For cats who have a natural propensity toward territoriality, or are given to fighting and spraying that do not respond to treatment, the choice may be between a life outdoors or no life at all.

Cats who tear up furniture and who cannot be broken of the habit fall into this category, too, though in their case there is the third alternative of declawing. Declawing is controversial among cat owners, and the Cat Fanciers' Association of America and some other organizations will not allow declawed cats to be shown in competitive events. Some veterinarians recommend against the practice on humane grounds and also point to anecdotal reports that declawed cats are more likely to develop behavioral problems. But a telephone survey of owners of clawed and declawed cats did not substantiate the latter claim, and other veterinarians point out that with the use of postsurgical painkillers the transient discomfort of the procedure can be minimized. Declawed cats are unable to defend themselves against dogs or other cats or climb trees very well to escape from threatening situations, and so do need to stay inside. But there is no evidence they suffer long-term impairment or pain; they jump and run as freely as clawed cats once they recover from the immediate effects of the surgery.

The Fate of the Cat

Yet ultimately, the goals of both the keep-all-cats-indoors zealots and the feed-and-protect-feral-cats zealots are probably unattainable. Many feral cats probably elude trapping, and indeed the net effect of trap-alter-release programs in the long run may be simply to create a powerful selective force in favor of an ever wilier and nastier population of feral cats, since those are the ones who will be left to reproduce.

The well-established role that farm cats play in limiting rodent populations in barns and fields, and the practical impossibility of keeping a certain fraction of the pet cat population indoors due to behavioral problems that show up under close confinement, likewise means that there will probably ever be a significant pool of free-roaming cats wherever human beings live. Only about 35 percent of cat owners in the United States keep their pets exclusively indoors, according to a 1997 poll; 31 percent let them outdoors some of the time and the remaining 34 percent keep them out all the time. There has always been population flow back and forth between the worlds of the pet cat and the worlds of the feral cat, and through the various demiworlds in between. Even today perhaps a third of pet cats are adopted from the free-living population, and conversely a significant fraction of feral cats are strays or abandoned pets. Even more striking is the statistic that only about a quarter of cat owners say they deliberately made an effort to acquire a cat: Their cats acquired them. Cats are now a global fact of nature, every bit as much as the tides that rise and fall each day, the barn swallows that return each spring, and, for that matter, the cockroach and the common cold. Human beings were the original vector for the cat's conquest of the world, but even today, four thousand years later, human beings have only a limited power or desire to control what they unwittingly unleashed. The ancient Egyptians were the sorcerer's apprentice of this tale; they knew enough to capture a wild animal and bring it up in a strange new world; they could tame the cat but not its destiny. A hundred coincidences in the makeup of *Felis silvestris* determined that once brought into the company of man the cat would take the world by storm, with scarcely a glance back, and certainly without a word of gratitude.

Cats are, now as then, an evolutionary force unto themselves. If that force is destined to drive the wildcat to extinction through genetic hybridization, send the balance of ecosystems of predators and prey tottering, spread disease, or provoke in us moral pain at the thought of the suffering of feral cats or the suffering of the birds and rodents they prey upon, the truth is there is probably little that people can do or will do to deflect that force in a different direction.

In some ways, though, we ought to admit that that's just as well. A cat who was totally a creature of human desires and ambitions would cease to be a cat. I don't think it is really the cat's stereotyped aloofness and independence that appeal to us, so much as his uncompromising otherness. The beauty and fascination we find in cats are much the same as what we feel for the wildest things in nature, with the added fascination that these particular wild and beautiful things are willing to admit us to their world, even though they don't have any particular need to. To be accepted by cats as an honorary cat is an honor indeed. But it is an honor we can only fully appreciate by seeing the world, if only in a glimpse, as the cat sees it. As the conservationist Aldo Leopold once observed, a true appreciation of what wild things mean to the human spirit can come only from understanding where they came from and how they lead their lives. And that is a story of science and understanding as much as it is one of spirit and empathy.

ACKNOWLEDGMENTS

I am grateful to Ellen Barber of the National Library of Medicine for her kind assistance, and to the many scientists who generously shared their knowledge with me—Carlos Driscoll, Marianne Hartmann, Peter Jackson, Andrew Kitchener, Marilyn Menotti-Raymond, Jay Neitz, and Stephen O'Brien in particular.

SOURCES AND FURTHER READING

1. CATS PLOT TO TAKE OVER THE WORLD, AND SUCCEED

Budiansky, Stephen. *The Covenant of the Wild: Why Animals Chose Domestication.* New York: Morrow, 1992; New Haven: Yale University Press, 1999.

Clutton-Brock, Juliet. *Domesticated Animals from Early Times.* Austin: University of Texas Press, 1981.

———. *Cats, Ancient and Modern.* Cambridge: Harvard University Press, 1993.

Davis, S. J. M. "Some More Animal Remains from the Aceramic Neolithic of Cyprus." In *Fouilles récentes à Khirokitia (Chypre), 1983–1986,* edited by Alain Le Brun. Paris: Editions Recherche sur les Civilisations, 1989.

Hubbard, A. L., et al. "Is Survival of European Wildcats *Felis silvestris* Threatened by Interbreeding with Domestic Cats?" *Biological Conservation* 61 (1992): 203–8.

Mackenzie, Donald Alexander. *Egyptian Myths and Legends.* 1910. Reprint. New York: Gramercy Books, 1994.

Masuda, Ryuichi, et al. "Molecular Phylogeny of Mitochondrial Cytochrome b and 12S rRNA Sequences in the Felidae: Ocelot and

Domestic Cat Lineages." *Molecular Phylogenetics and Evolution* 6 (1996): 351–65.

Nowell, Kristin, and Peter Jackson, eds. *Wild Cats: Status Survey and Conservation Action Plan.* Gland, Switzerland: IUCN, 1996.

O'Brien, Stephen J., Johannes Wienberg, and Leslie A. Lyons. "Comparative Genomics: Lessons from Cats." *Trends in Genetics* 10 (1997): 393–99.

O'Brien, Stephen J., et al. "Comparative Gene Mapping in the Domestic Cat (*Felis catus*)." *Journal of Heredity* 88 (1997): 408–14.

Randi, E., and B. Ragni. "Genetic Variability and Biochemical Systematics of Domestic and Wild Cat Populations (*Felis silvestris:* Felidae)." *Journal of Mammalogy* 72 (1991): 79–88.

Serpell, James A. "The Domestication and History of the Cat." In *The Domestic Cat: The Biology of Its Behavior,* edited by Dennis C. Turner and Patrick Bateson. Cambridge: Cambridge University Press, 1988.

Smithers, Reay H. N. "Cat of the Pharaohs." *Animal Kingdom* 71 (1968): 16–23.

Stahl, P., and M. Artois. *Status and Conservation of the Wildcat in Europe and Around the Mediterranean Rim.* Council of Europe Nature and Environment Series No. 69, 1994.

Williams, Robert W., Carmen Cavada, and Fernando Reinoso-Suárez. "Rapid Evolution of the Visual System: A Cellular Assay of the Retina and Dorsal Lateral Geniculate Nucleus of the Spanish Wildcat and the Domestic Cat." *Journal of Neuroscience* 13 (1993): 208–28.

2. BLACK CATS AND TABBY CATS

Blumenberg, B., and A. T. Lloyd. "Mutant Allele Frequencies in the Domestic Cat: A Preliminary Discussion of Selection with Particular Reference to the United Kingdom and Eire." *Genetica* 54 (1980): 17–28.

Clark, J. M. "The Effects of Selection and Human Preference on Coat Color Gene Frequencies in Urban Cats." *Heredity* 35 (1975): 195–210.

Dale-Green, Patricia. *Cult of the Cat.* Boston: Houghton Mifflin, 1963.

Darnton, Robert. *The Great Cat Massacre and Other Episodes in French Cultural History.* New York: Basic Books, 1984.

Howey, M. Oldfield. *The Cat in the Mysteries of Religion and Magic.* London: Rider & Co., 1930.

Lomax, T. D., and R. Robinson. "Tabby Pattern Alleles of the Domestic Cat." *Journal of Heredity* 79 (1988): 21–23.

Menotti-Raymond, Marilyn A., Victor A. David, and Stephen J. O'Brien. "Pet Cat Hair Implicates Murder Suspect." *Nature* 386 (1997): 774.

Méry, Fernand. *The Life, History, and Magic of the Cat,* translated by Emma Street. New York: Madison Square Press, 1968.

Miller, Joan. "The Domestic Cat: Perspective on the Nature and Diversity of Cats." *Journal of the American Veterinary Medical Association* 208 (1996): 498–502.

O'Brien, Stephen J. "The Extent and Character of Biochemical Genetic Variation in the Domestic Cat." *Journal of Heredity* 71 (1980): 2–8.

Todd, Neil B. "Cats and Commerce." *Scientific American* November 1977: 101–7.

Vella, Carolyn M., et al. *Robinson's Genetics for Cat Breeders and Veterinarians.* 4th edition. Oxford: Butterworth-Heinemann, 1999.

Vinogradov, Alexander E. "Locally Associated Alleles of Cat Coat Genes." *Journal of Heredity* 85 (1994): 86–91.

———. "Fine Structure of Gene Frequency Landscapes in Domestic Cat: The Old and New Worlds Compared." *Hereditas* 126 (1997): 95–102.

3. THE WAR BETWEEN THE SEXES AND OTHER ODDITIES OF FELINE SOCIETY

Apps, Peter James. "Home Ranges of Feral Cats on Dassen Island." *Journal of Mammalogy* 67 (1986): 199–200.

Bradshaw, John W. S., and Sarah L. Brown. "Social Behaviour of Cats." *Tijdschrift voor Diergeneeskunde* 117, suppl. 1 (1992): 54S–56S.

Crowell-David, Sharon L., Kimberly Barry, and Randall Wolfe. "Social Behavior and Aggressive Problems in Cats." *Veterinary Clinics of North America: Small Animal Practice* 27 (1997): 549–67.

Diakow, Carol. "Effects of Genital Densensitization on Mating Behavior and Ovulation in the Female Cat." *Physiology & Behavior* 7 (1971): 47–54.

Durr, Rena, and Christopher Smith. "Individual Differences and Their Relation to Social Structure in Domestic Cats." *Journal of Comparative Psychology* 111 (1997): 412–18.

Feldman, Hilary N. "Maternal Care and Differences in the Use of Nests in the Domestic Cat." *Animal Behaviour* 45 (1993): 13–23.

Hart, B. L., and R. E. Barrett. "Effects of Castration on Fighting, Roaming, and Urine Spraying in Adult Male Cats." *Journal of the American Veterinary Medical Association* 163 (1973): 290.

Hendriks, Wouter H., Michael F. Tarttelin, and Paul J. Moughan. "Twenty-four Hour Feline Excretion Patterns in Entire and Castrated Cats." *Physiology & Behavior* 58 (1995): 467–69.

Kerby, G., and D. W. Macdonald. "Cat Society and the Consequences of Colony Size." In *The Domestic Cat: The Biology of Its Behavior,* edited by Dennis C. Turner and Patrick Bateson. Cambridge: Cambridge University Press, 1988.

Laundré, John. "The Daytime Behaviour of Domestic Cats in a Free-Roaming Population." *Animal Behavior* 25 (1977): 990–98.

Leyhausen, Paul. "The Tame and the Wild: Another Just-So Story?" In *The Domestic Cat: The Biology of Its Behavior,* edited by Dennis C. Turner and Patrick Bateson. Cambridge: Cambridge University Press, 1988.

Liberg, Olof, and Mikael Sandell. "Spatial Organization and Reproductive Tactics in the Domestic Cat and Other Felids." In *The Domestic Cat: The Biology of Its Behavior,* edited by Dennis C. Turner and Patrick Bateson. Cambridge: Cambridge University Press, 1988.

Martin, Paul. "The Time and Energy Costs of Play Behaviour in the Cat." *Zeitschrift für Tierpsychologie* 64 (1984): 298–312.

Natoli, Eugenia. "Behavioural Responses of Urban Feral Cats to Different Types of Urine Marks." *Behaviour* 94 (1985): 234–43.

———. "Spacing Pattern in a Colony of Urban Stray Cats (*Felis catus* L.)

in the Historic Centre of Rome." *Applied Animal Behaviour Science* 14 (1985): 289–304.

———. "Mating Strategies in Cats: A Comparison of the Role and Importance of Infanticide in Domestic Cats, *Felis catus* L., and Lions, *Panthera leo* L." *Animal Behaviour* 40 (1990): 183–86.

Natoli, Eugenia, and Emanuele De Vito. "The Mating System of Feral Cats Living in a Group." In *The Domestic Cat: The Biology of Its Behavior,* edited by Dennis C. Turner and Patrick Bateson. Cambridge: Cambridge University Press, 1988.

———. "Agonistic Behaviour, Dominance Rank and Copulatory Success in a Large Multi-male Feral Cat, *Felis catus* L., Colony in Central Rome." *Animal Behaviour* 42 (1991): 227–41.

Panaman, Roger. "Behavior and Ecology of Free-Ranging Female Farm Cats (*Felis catus* L.)" *Zeitschrift für Tierpsychologie* 56 (1981): 59–73.

West, Meredith. "Social Play in the Domestic Cat." *American Zoologist* 14 (1974): 427–36.

Wolski, Thomas R. "Social Behavior of the Cat." *Veterinary Clinics of North America: Small Animal Practice* 12 (1982): 693–706.

4. OUTTA MY FACE, AND OTHER USEFUL EXPRESSIONS

Baron, Alan, C. N. Stewart, and J. M. Warren. "Patterns of Social Interaction in Cats (*Felis domestica*)." *Behaviour* 11 (1957): 56–66.

Beaver, Bonnie V. "The Marking Behavior of Cats." *Veterinary Medicine/Small Animal Clinician* 76 (1981): 792–93.

Cole, D. D., and J. N. Shafer. "A Study of Social Dominance in Cats." *Behaviour* 27 (1966): 39–53.

De Boer, J. N. "Dominance Relations in Pairs of Domestic Cats." *Behavioural Processes* 2 (1977): 227–42.

Feldman, Hilary N. "Domestic Cats and Passive Submission." *Animal Behaviour* 47 (1994): 457–59.

Freeman, Natalie C. G., and Jay S. Rosenblatt. "The Interrelationship Between Thermal and Olfactory Stimulation in the Development of Home Orientation in Newborn Kittens." *Developmental Psychobiology* 11 (1978): 437–57.

———. "Specificity of Litter Odors in the Control of Home Orientation Among Kittens." *Developmental Psychobiology* 11 (1978): 459–68.

Haskins, Ron. "Effect of Kitten Vocalizations on Maternal Behavior." *Journal of Comparative and Physiological Psychology* 91 (1977): 830–38.

———. "A Causal Analysis of Kitten Vocalization: An Observational and Experimental Study." *Animal Behaviour* 27 (1979): 726–36.

Houpt, Katherine A. *Domestic Animal Behavior for Veterinarians and Animal Scientists.* 3rd edition. Ames: Iowa State University Press, 1998.

Kiley-Worthington, M. "The Tail Movements of Ungulates, Canids, and Felids with Particular Reference to Their Causation and Function as Displays." *Behaviour* 56 (1976): 69–115.

Owings, Donald H., and Eugene S. Morton. *Animal Vocal Communication: A New Approach.* Cambridge: Cambridge University Press, 1998.

Remmers, J. E., and H. Gautier. "Neural and Mechanical Mechanisms of Feline Purring." *Respiration Physiology* 16 (1972): 351–61.

Wolski, Thomas R. "Social Behavior of the Cat." *Veterinary Clinics of North America: Small Animal Practice* 12 (1982): 693–706.

5. THE THINKING CAT'S GUIDE TO INTELLIGENCE

Adamec, Robert E. "The Interaction of Hunger and Prey in the Domestic Cat (*Felis catus*): An Adaptive Hierarchy?" *Behavioral Biology* 18 (1976): 263–72.

Blake, Randolph. "Cats Perceive Biological Motion." *Psychological Science* 4 (1993): 54–57.

Blake, Randolph, and William Martens. "Critical Bands in Cat Spatial Vision." *Journal of Physiology* 314 (1981): 175–87.

Caro, T. M., and M. D. Hauser. "Is There Teaching in Nonhuman Animals?" *Quarterly Review of Biology* 67 (1992): 151–74.

Chesler, Phyllis. "Maternal Influence in Learning by Observation in Kittens." *Science* 166 (1969): 901–3.

Collier, George, Deanne F. Johnson, and Cynthia Morgan. "Meal Patterns of Cats Encountering Variable Food Procurement Costs." *Journal of the Experimental Analysis of Behavior* 67 (1997): 303–10.

Davis, Joel L., and Robert A. Jensen. "The Development of Passive and Active Avoidance Learning in the Cat." *Developmental Psychobiology* 9 (1976): 175–79.

Doré, François Y. "Search Behavior of Cats (*Felis catus*) in an Invisible Displacement Test: Cognition and Experience." *Canadian Journal of Psychology* 44 (1990): 359–70.

Dumas, Claude. "Object Permanence in Cats (*Felis catus*): An Ecological Approach to the Study of Invisible Displacements." *Journal of Comparative Psychology* 106 (1992): 404–10.

Dumas, Claude, and François Y. Doré. "Cognitive Development in Kittens (*Felis catus*): An Observational Study of Object Permanence and Sensorimotor Intelligence." *Journal of Comparative Psychology* 105 (1991): 357–65.

Fiset, Sylvain, and François Y. Doré. "Spatial Encoding in Domestic Cats (*Felis catus*)." *Journal of Experimental Psychology: Animal Behavior Processes* 22 (1996): 420–37.

Harrison, Jean, and Jennifer Buchwald. "Eyeblink Conditioning Deficits in the Old Cat." *Neurobiology of Aging* 4 (1983): 45–51.

Hart, Benjamin L. "Learning Ability in Cats." *Feline Practice* 5(5): 10–12 (September–October 1975).

Heishman, Miriam, Mindy Conant, and Robert Pasnak. "Human Analog Tests of the Sixth Stage of Object Permanence." *Perceptual and Motor Skills* 80 (1995): 1059–68.

John, E. R., et al. "Observation Learning in Cats." *Science* 159 (1968): 1589–91.

Kukorelli, Tibor, and Laszlo Detari. "Effects of Viscerosensory Stimulation on Hypothalamically Elicited Predatory Behavior in Cats." *Physiology & Behavior* 55 (1994): 705–10.

Leveque, Nancy W. "Cognitive Dysfunction in Dogs, Cats an Alzheimer's-like Disease." *Journal of the American Veterinary Medical Association* 212 (1998): 1351.

Loop, Michael S., C. Leigh Millican, and Shari R. Thomas. "Photopic Spectral Sensitivity of the Cat." *Journal of Physiology* 382 (1987): 537–53.

Martin, Paul, and Patrick Bateson. "Behavioural Development in the Cat." In *The Domestic Cat: The Biology of Its Behavior,* edited by Dennis C. Turner and Patrick Bateson. Cambridge: Cambridge University Press, 1988.

Oniani, T. N., N. D. Lortkipanidze, and L. M. Maisuradze. "Interaction Between Learning and Paradoxical Sleep in Cats." *Neuroscience and Behavioral Physiology* 17 (1987): 304–10.

Perfiliev, S., L. G. Pettersson, and A. Lundberg. "Control of Claw Movements in Cats." *Neuroscience Research* 31 (1998): 337–42.

Rosenblatt, Jay S. "Suckling and Home Orientation in the Kitten: A Comparative Developmental Study." In *The Biopsychology of Development,* edited by Ethel Tobach, Lester R. Aronson, and Evelyn Shaw. New York: Academic Press, 1971.

Rosenkilde, Carl E., and Ivan Divac. "Discrimination of Time Intervals in Cats." *Acta Neurobiologiae Experimentalis* 36 (1976): 311–17.

Sherk, Helen, and Garth A. Fowler. "Optic Flow and the Visual Guidance of Locomotion in the Cat." *International Review of Neurobiology* 44 (2000): 141–70.

Voith, Victoria L. "You, Too, Can Teach a Cat Tricks (Examples of Shaping, Second-Order Reinforcement, and Constraints on Learning)." *Modern Veterinary Practice,* August 1981: 639–42.

Warren, J. M., and Alan Barron. "The Formation of Learning Sets by Cats." *Journal of Comparative and Physiological Psychology* 49 (1956): 227–31.

Zernicki, Boguslaw. "Effects of Binocular Deprivation and Specific Experience in Cats: Behavioral, Electrophysiological, and Biochemical Analyses." In *Brain Mechanisms in Memory and Learning: From the Single Neuron to Man,* edited by M. A. B. Brazier. New York: Raven Press, 1979.

6. THE CAT PERSONALITY TEST

Adamec, Robert E. "Individual Differences in Temporal Lobe Sensory Processing of Threatening Stimuli in the Cat." *Physiology & Behavior* 49 (1991): 455–64.

Candland, Douglas K., and David Milne. "Species Differences in Approach-Behaviour as a Function of Developmental Environment." *Animal Behaviour* 14 (1966): 539–45.

Caro, T. M. "The Effects of Experience on the Predatory Patterns of Cats." *Behavioral and Neural Biology* 29 (1980): 1–28.

———. "Effects of the Mother, Object Play, and Adult Experience on Predation in Cats." *Behavioral and Neural Biology* 29 (1980): 28–51.

Feaver, Julie, Michael Mendl, and Patrick Bateson. "A Method for Rating the Individual Distinctiveness of Domestic Cats." *Animal Behaviour* 34 (1986): 1016–25.

Gallo, Patricia Vetula, Jack Werboff, and Kirvin Knox. "Development of Home Orientation in Offspring of Protein-Restricted Cats." *Developmental Psychobiology* 17 (1984): 437–49.

Hart, Benjamin L., and Mitzi G. Leedy. "Analysis of the Catnip Reaction: Mediation by Olfactory System, Not Vomeronasal Organ." *Behavioral and Neural Biology* 44 (1985): 38–46.

Hill, J. O., et al. "Species-Characteristic Responses to Catnip by Undomesticated Felids." *Journal of Chemical Ecology* 2 (1976): 239–53.

Karsh, Eileen B., and Dennis C. Turner. "The Human–Cat Relationship." In *The Domestic Cat: The Biology of Its Behavior*, edited by Dennis C. Turner and Patrick Bateson. Cambridge: Cambridge University Press, 1988.

Kolb, Bryan, and Arthur J. Nonneman. "The Development of Social Responsiveness in Kittens." *Animal Behaviour* 23 (1975): 368–74.

Konrad, Karl W., and Muriel Bagshaw. "Effect of Novel Stimuli on Cats Reared in a Restricted Environment." *Journal of Comparative and Physiological Psychology* 70 (1970): 157–64.

Martin, Paul, and Patrick Bateson. "The Ontogeny of Locomotor Play Behaviour in the Domestic Cat." *Animal Behaviour* 33 (1985): 502–11.

———. "The Influence of Experimentally Manipulating a Component of Weaning on the Development of Play in Domestic Cats." *Animal Behaviour* 33 (1985): 511–18.

Meier, Gilbert W. "Infantile Handling and Development in Siamese Kittens." *Journal of Comparative and Physiological Psychology* 54 (1961): 284–86.

Meier, Gilbert W., and Jane L. Stuart. "Effects of Handling on the Physical and Behavioral Development of Siamese Kittens." *Psychological Reports* 5 (1959): 497–501.

Paré, Denis, and Dawn R. Collins. "Neuronal Correlates of Fear in the Lateral Amygdala: Multiple Extracellular Recordings in Conscious Cats." *Journal of Neuroscience* 20 (2000): 2701–10.

Podberscek, A. L., J. K. Blackshaw, and A. W. Beattie. "The Behavior of Laboratory Colony Cats and Their Reactions to a Familiar and Unfamiliar Person." *Applied Animal Behaviour Science* 31 (1992): 119–30.

Seitz, Philip F. D. "Infantile Experience and Adult Behavior in Animal Subjects. II. Age of Separation from the Mother and Adult Behavior in the Cat." *Psychosomatic Medicine* 21 (1959): 353–78.

Shaikh, Majid B. et al. "Dopaminergic Regulation of Quiet Biting Attack Behavior in the Cat." *Brain Research Bulletin* 27 (1991): 725–30.

Siegel, Allan, and Kristie Schubert. "Neurotransmitters Regulating Feline Aggressive Behavior." *Reviews in the Neurosciences* 6 (1995): 47–61.

Siegel, Allan, Kristie L. Schubert, and Majid B. Shaikh. "Neurotransmitters Regulating Defensive Rage Behavior in the Cat." *Neuroscience and Biobehavioral Reviews* 21 (1997): 733–42.

Wilson, Margaret, J. M. Warren, and Lynn Abbott. "Infantile Stimulation, Activity, and Learning by Cats." *Child Development* 36 (1965): 843–53.

Wyrwicka, Wanda. "Social Effects on Development of Food Preferences." *Acta Neurobiologiae Experimentalis* 53 (1993): 485–93.

Zernicki, Boguslaw. "Learning Deficits in Lab-Reared Cats." *Acta Neurobiologiae Experimentalis* 53 (1993): 231–36.

7. CATS AND TROUBLE

Adamec, Robert E. "Anxious Personality in the Cat: Its Ontogeny and Physiology." In *Psychopathology and the Brain*, edited by Bernard J.

Carroll and James E. Barrett. New York: Raven Press, 1991.

Chapman, Barbara L. "Feline Aggression. Classification, Diagnosis, and Treatment." *Veterinary Clinics of North America: Small Animal Practice* 21 (1991): 315–27.

Cooper, Leslie Larson. "Feline Inappropriate Elimination." *Veterinary Clinics of North America: Small Animal Practice* 27 (1997): 569–600.

Crowell-David, Sharon L., Kimberly Barry, and Randall Wolfe. "Social Behavior and Aggressive Problems in Cats." *Veterinary Clinics of North America: Small Animal Practice* 27 (1997): 549–67.

Edenburg, N., and B. W. Knol. "Behavioural, Household, and Social Problems Associated with Companion Animals: Opinions of Owners and Non-owners." *Veterinary Quarterly* 16 (1994): 130–34.

Hart, Benjamin L. "Behavioral Therapy with Mousetraps." *Feline Practice* 9(4): 10–14 (July–August 1979).

———. "Starting from Scratch: New Perspectives on Cat Scratching." *Feline Practice* 10(4): 8–12 (July–August 1980).

———. "Behavioral and Pharmacological Approaches to Problem Urination in Cats." *Veterinary Clinics of North America: Small Animal Practice* 26 (1996): 651–58.

Hart, Benjamin L., and R. E. Barrett. "Effects of Castration on Fighting, Roaming, and Urine Spraying in Adult Male Cats." *Journal of the American Veterinary Medical Association* 163 (1973): 290.

Hart, Benjamin L., and Lynette A. Hart. *Canine and Feline Behavioral Therapy.* Philadelphia: Lea & Febiger, 1985.

Hart, Benjamin L., et al. "Effectiveness of Buspirone on Urine Spraying and Inappropriate Urination in Cats." *Journal of the American Veterinary Medical Association* 203 (1993): 254–58.

Houpt, Katherine A., Sue Utter Honig, and Ilana R. Reisner. "Breaking the Human–Companion Animal Bond." *Journal of the American Veterinary Medical Association* 208 (1996): 1653–59.

Landsberg, Gary. "Feline Behavior and Welfare." *Journal of the American Veterinary Medical Association* 208 (1996): 502–5.

Morgan, Melanie, and Katherine A. Houpt. "Feline Behavior Problems:

The Influence of Declawing." *Anthrozoös* 3 (1989): 50–53.

Olm, Dale D., and Katherine A. Houpt. "Feline House-Soiling Problems." *Applied Animal Behaviour Science* 20 (1988): 335–45.

Patrick, Gail R., and Kathleen M. O'Rourke. "Dog and Cat Bites: Epidemiologic Analyses Suggest Different Prevention Strategies." *Public Health Reports* 113 (1998): 252–57.

Patronek, Gary J. "Free-Roaming and Feral Cats—Their Impact on Wildlife and Human Beings." *Journal of the American Veterinary Medical Association* 212 (1998): 218–26.

Robertson, I.D. "Survey of Predation by Domestic Cats." *Australian Veterinary Journal* 76 (1998): 551–54.

Schwartz, Stefanie. "Use of Cyproheptadine to Control Urine Spraying and Masturbation in a Cat." *Journal of the American Veterinary Medical Association* 214 (1999): 369–71.

———. "Use of Cyproheptadine to Control Urine Spraying in a Castrated Male Cat." *Journal of the American Veterinary Medical Association* 215 (1999): 501–2.

Seksel, S., and M. J. Lindeman. "Use of Clomipramine in the Treatment of Anxiety-Related and Obsessive-Compulsive Disorders in Cats." *Australian Veterinary Journal* 76 (1998): 317–21.

Sterman, M. B., et al. "Circadian Sleep and Waking Patterns in the Laboratory Cat." *Electroencephalography and Clinical Neurophysiology* 19 (1965): 509–17.

Stubbs, W. Preston, et al. "Effects of Prepubertal Gonadectomy on Physical and Behavioral Development in Cats." *Journal of the American Veterinary Medical Association* 209 (1996): 1864–71.

Willemse, Ton, et al. "The Effect of Haloperidol and Naloxone on Excessive Grooming Behavior of Cats." *European Neuropsychopharmacology* 4 (1994): 39–45.

ILLUSTRATION CREDITS

PLATES

I. From John Anderson, *Zoology of Egypt: Mammalia* (London: B. Quaritch, 1902).
II. Egyptian expedition of The Metropolitan Museum of Art, Rogers Fund, 1930 (30.4.114). Photograph © 1979 The Metropolitan Museum of Art.
III. © The British Museum.
IV. Illustration by Enid Kotschnig from Neil Todd, "Cats and Commerce," *Scientific American*, November 1977.

FIGURES

page

7. Illustration by Dave Merrill; basic data from Stephen O'Brien.
12. Illustration by Dave Merrill; basic data from The World Conservation Union.
24. Illustration by Dave Merrill.
25, 26. © The British Museum.
43. From *The Wonderful Discoverie of the Witchcrafts of Margaret and Phillip[pa] Flower* (London, 1619).
51, 52, 59. Illustrations by Dave Merrill; basic data from Neil Todd, "Cats and Commerce," *Scientific American*, November 1977.
84. Illustration by Dave Merrill; basic data from Priscilla Barrett and Patrick Bateson, "The Development of Play in Cats," *Behaviour* 66 (1978): 106–20.
96, 97. Illustrations by Dave Merrill; basic data from Paul Leyhausen, *Cat Behavior: The Predatory and Social Behavior of Domestic and Wild Cats* (New York: Garland Press, 1979).
112. Illustration by Dave Merrill.
123. Illustration by Dave Merrill; basic data from Kirk N. Gelatt, ed., *Veterinary Ophthalmology*, 3rd edition (Philadelphia: Lippincott, Williams & Wilkins, 1999).
124. Illustration by Dave Merrill.
130. Animals Animals, © Gerald Lucz.
193. Illustration by Dave Merrill; basic data from M. B. Stearman et al., "Circadian Sleep and Waking Patterns in the Laboratory Cat," *Electroencephalography and Clinical Neurophysiology* 19 (1965): 509–17.

INDEX